# STUDENT UNIT GUIDE

## NEW EDITION

Edexcel A2 Geography Unit 4

Geographical Research

Kim Adams, David Holmes
and Cameron Dunn

Philip Allan Updates, an imprint of Hodder Education, an Hachette UK company, Market Place, Deddington, Oxfordshire OX15 0SE

*Orders*
Bookpoint Ltd, 130 Milton Park, Abingdon, Oxfordshire OX14 4SB
tel: 01235 827827
fax: 01235 400401
e-mail: education@bookpoint.co.uk
Lines are open 9.00 a.m.–5.00 p.m., Monday to Saturday, with a 24-hour message answering service.
You can also order through the Philip Allan Updates website: www.philipallan.co.uk

ISBN 978-1-4441-4770-4

First printed 2012
Impression number 5 4 3 2 1
Year 2016 2015 2014 2013 2012

Printed in Dubai

Hachette UK's policy is to use papers that are natural, renewable and recyclable products and made from wood grown in sustainable forests. The logging and manufacturing processes are expected to conform to the environmental regulations of the country of origin.

P01987

# Contents

# Getting the most from this book

## Examiner tips

Advice from the examiner on key points in the text to help you learn and recall unit content, avoid pitfalls, and polish your exam technique in order to boost your grade.

## Knowledge check

Rapid-fire questions throughout the Content Guidance section to check your understanding.

## Knowledge check answers

**1** Turn to the back of the book for the Knowledge check answers.

## Summary

Summaries

● Each core topic is rounded off by a bullet-list summary for quick-check reference of what you need to know.

## Questions & Answers

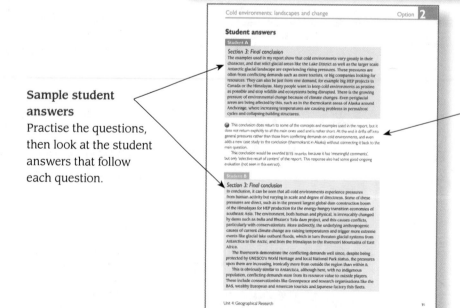

**Sample student answers**
Practise the questions, then look at the student answers that follow each question.

**Examiner commentary on sample student answers**
Find out how many marks each answer would be awarded in the exam and then read the examiner comments (preceded by the icon 🄴) following each student answer.

# About this book

This guide is for students following the Edexcel A2 Geography course. It aims to guide you through Unit 4: Geographical Research.

This introduction provides an overview of the unit, introduces the specification content and gives tips on carrying out research.

The **Content Guidance** section provides a detailed guide to the six options from which you will choose one to specialise in. Research activities are also given.

The **Questions & Answers** section provides exam guidance and samples of different grades of work in the final report with examiner's comments on how to improve your performance in the exam.

## Overview of Unit 4

For Unit 4: Geographical Research you must choose **one** from six available options. In the exam you have to write one report in 1.5 hours. Some options are more physically biased, some more human.

The options (starting with the more physically based options and moving towards the more human based options) are as follows:

Option 1 Tectonic activity and hazards

Option 2 Cold environments: landscapes and change

Option 3 Life on the margins: the food supply problem

Option 4 The world of cultural diversity

Option 5 Pollution and human health at risk

Option 6 Consuming the rural landscape: leisure and tourism

In the specification, each option is divided into four enquiry questions, and these are then divided into four sub-enquiry topics. An example based on Option 2 is shown in Table A.

**Table A** Cold environments: landscapes and change

| Main enquiry questions | 1 Defining and locating cold environments | 2 Climatic processes and their causes | 3 Distinctive landforms and landscapes | 4 Challenges and opportunities |
|---|---|---|---|---|
| **Sub-enquiry topics** | • Cold, glacial, periglacial environments<br>• Systems<br>• Distribution changes over time<br>• British Isles glacial and interglacial history | • Climatic causes<br>• Long-term changes<br>• Meteorological processes<br>• British Isles spatial and temporal relationships between glacial and periglacial environments | • Glacial geomorphological processes<br>• Glacial landforms<br>• Periglacial geomorphological processes<br>• Periglacial landforms | • Definitions and links<br>• Past and present challenges and opportunities<br>• Overcoming challenges<br>• Approaches to use and management |

Questions in the final exam may be based mainly on one of the four enquiry questions, but are more likely to be based across several. Your work on this unit may be thought of as a progression along a **timeline** (see Table B). The important thing throughout this sustained piece of work is to create an orderly set of notes and annotated articles so that when you are given the research focus for the real exam a few weeks in advance, you have identifiable sections to concentrate on, a trail of extra references/ sources to check concepts and improve factual details, and, if relevant, additional topical examples.

**Table B** Timeline

| | |
|---|---|
| **Decision on which option**<br><br>Approx 10 weeks' class work and homework, researching into each of the sub-enquiry questions for ONE option | • Your notes should make clear reference to the main four enquiry questions. Build up a series of detailed case studies and smaller examples so that you can be flexible in the final exam. There is as much skill in rejecting information that you know as there is in selecting the most appropriate bits.<br>• Summarise information in factfiles, glossaries, spider topic webs.<br>• Each example you research should have information on the key synoptic elements of this unit: people/places/power.<br>• Before the final exam you should write at least five practice reports, some under timed conditions, with/without your notes. Create a glossary of key terms and learn them. |
| **The pre-release research focus, given at least 4 working weeks before the actual exam** | • Look closely at the research focus, it gives major clues as to the final title in the exam.<br>• Practise different command words, such as:<br>  – Discuss…<br>  – To what extent…<br>  – Explain why or how…<br>• Ensure you understand key terms like process, factor, players, geographical, challenges, responses, causes, impacts.<br>• Make short summaries on cards of key facts to learn for the exam.<br>• Learn your glossary – geographical vocabulary is essential. |
| **The exam**<br>90 minutes | • You will have a page in your exam booklet devoted to making a plan – use it. It doesn't matter how messy and scribbled this is, it's for you not the examiner. Students who spend 5 minutes or so on a plan usually increase their final marks substantially.<br>• Then get writing, remembering the outline of how your report will be marked (see Exam tips: DRACQ).<br>• Try to finish the report – even at the risk of making it obvious that you ran out of time by leaving a gap and going to the conclusion. Otherwise you throw away valuable marks. |

# Content guidance

# Tectonic activity and hazards

## Introduction

Hazards, including tectonic activity, become disasters because of human vulnerability. An earthquake in an unpopulated area is not a disaster, because there is no human impact. Tectonic activity generates a wide range of natural hazards including lava flows, ash falls, ground shaking and some notable tsunamis (e.g. Japan 2011). The primary cause of these is plate tectonics — the movement of the crustal plates on the Earth's surface. As a result, tectonic hazards have a particular distribution concentrated on the plate margins and 'hot spots'.

Tectonic hazards can generate significant risk (short, medium and longer term) to people who live in tectonically active regions. Risks are closely related to:

- vulnerability of the population
- magnitude and frequency of hazard events
- level of economic development
- population density
- community preparedness/education of the people

People respond to hazard risks in a variety of ways, e.g. trying to prevent or modify the hazardous event, modifying the loss burden, and reducing human vulnerability. The level, scale and effectiveness of the response depend on available resources. Much current scientific research is focused on increasing our understanding of tectonic hazards and on improving our ability to cope with their impacts.

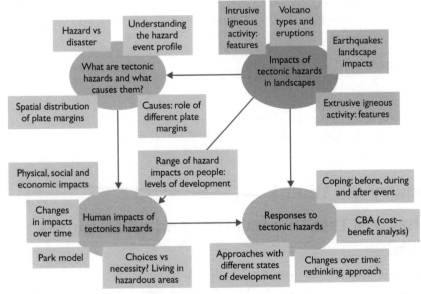

**Figure 1** Topic concept map

While thinking of the synoptic aspects of the course (Figure 2), a simple checklist based on the topics within each enquiry question in the specification will help you to separate your research into manageable chunks. When you are researching tectonics, use a checklist linked to the specification (see Table 1). Remember to note the main reference source as this is a requirement for the exam — you are expected to quote some key sources: from books, newspapers, websites, DVDs, podcasts, pamphlets etc.

Describe the usual relationships between magnitude, level of development and vulnerability.

**Figure 2** The synoptic context

**Table 1** Checklist for tectonics

| Enquiry question 1<br><br>What are tectonic hazards and what causes them? | Enquiry question 2<br><br>What impact does tectonic activity have on landscapes and why does this impact vary? | Enquiry question 3<br><br>What impacts do tectonic hazards have on people and why do these impacts vary? | Enquiry question 4<br><br>How do people cope with tectonic hazards and what are the issues for the future? |
|---|---|---|---|
| • What is a tectonic event?<br>• Hazard vs disaster<br>• Global plate boundaries and types (transform, convergent etc.)<br>• Patterns and distribution — comparisons of earthquakes vs volcanoes<br>• Understanding the hazard profile (and key terms) | • How extrusive igneous activity varies spatially<br>• Features linked to intrusive igneous activity<br>• Formation and types of volcano<br>• Plate boundaries vs hot spots<br>• Earthquake impacts on landscapes and associated features | • Tectonic risk and people — influences<br>• Perceived vs real tectonic risks<br>• Comparison of risks in locations with different states of development<br>• Impacts on people: social and economic<br>• Hazards impacts are dynamic: long and shorter term changes | • Approaches to management: variations by place and level of economic development<br>• Modifying event?<br>• Modifying loss burden?<br>• Modifying vulnerability?<br>• Changes in approaches over time (before, during and after) |
| Key refs used | Key refs used | Key refs used | Key refs used |

Use the BBC website to keep a close eye on tectonic hazards as they happen. It's a good source of up-to-date information. For a more reflective digest of events, *The Economist* can be excellent, especially in terms of economic impacts.

Use Table 1 as an initial stimulus, and carry out your own research audit to obtain background sources of information. Make a note of any key references used.

# Enquiry question 1

# What are tectonic hazards and what causes them?

A **tectonic event** is a physical occurrence resulting from the movement or deformation of the Earth's crust. They are predominantly earthquakes or volcanic eruptions and their associated activities. Such events become **tectonic hazards** when they have the potential to cause loss of life and damage to property. But not all events are hazardous. Many earthquakes in active areas are low in magnitude and may occur deep below the surface. People living in the area may not even feel them. For example, in a typical week in April 2011 there were over 150 'earthquakes' in the area around Los Angeles, but virtually all were below 3.0 on the Richter scale. They presented no danger to the local people.

A hazard becomes a **tectonic disaster** when an event materialises and causes extensive destruction and a number of fatalities. For example, the famous historic eruption of Mt Pinatubo in the Philippines in 1991 killed about 300 people, over 80,000 homes were destroyed or damaged and 800 km² of agricultural land was buried beneath a thick layer of ash.

The **hazard event profile** is a diagram that tries to represent the characteristics (magnitude, speed of onset, duration, areal extent etc.) of different types of hazards (Figure 3).

**Knowledge check 2**

Know how to apply a hazard event profile to different types of event. What are the advantages of these diagrams?

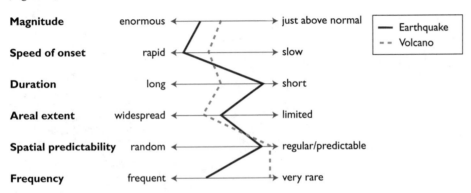

**Figure 3** Typical hazard event profile for an earthquake (destructive boundary) and volcano

**Examiner tip**

Be careful to get your information as up to date as possible, using data from the year of the exam you will be taking if possible. Remember to evaluate the quality of such sources, especially commenting on their reliability.

## Causes of tectonic hazards

The tectonic plates of the Earth's crust move relative to each other so there is a slow build up of stress within rocks. When the pressure is released, parts of the Earth's surface experience an intense shaking motion, which typically only lasts a few seconds. This is the main cause of earthquakes. Earthquakes can be classified according to the depth of their focus, which affects the amount of surface damage that results from the event. Three broad categories are recognised:

- deep focus: 300–700 km deep
- intermediate focus: 70–300 km deep
- shallow focus 0–70 km deep. These cause the greatest degree of damage and account for approximately 75% of all earthquake types

A volcano is a landform that develops around a weakness in the Earth's crust. At this point, molten rock or magna is forced out or extruded. The nature of the volcanic eruption, i.e. viscosity (thickness and stickiness of the magma), the amount of dissolved gases and how easily they can escape, is determined by the tectonic context. Volcanic eruptions can be classified by the 'volcanic explosivity index' (VEI) where 0 = non-explosive and 8 is very large.

**Table 2** Comparison of the distribution of volcanoes and earthquakes

| Earthquakes | Volcanoes |
|---|---|
| Distribution tends to be tightly focused in linear bands (belts), but can occasionally occur away from these, e.g. New Madrid earthquakes (USA) in 1811–12 and Dudley (UK) earthquake 2002 (both examples of an intra-plate earthquake) | A good proportion of volcanic activity occurs at or near plate margins (inter-plate). Can be at a constructive margin: e.g. Iceland, or destructive: e.g. Andean volcanoes in Chile and Peru. Generally distribution is both *linear* and *clustered* |
| Most intense and frequent earthquakes are associated with subduction/destructive zones, and sometimes active conservative margins, e.g. San Andreas, USA | Volcanoes can also be found intra-plate at hot spots, e.g. Hawaii in the Pacific |
| High intensity shake events are *not* normally associated with non-active constructive (transform) boundaries | Continental margins are associated with the most violent volcanic eruptions (due to the presence of both oceanic and continental crust) |

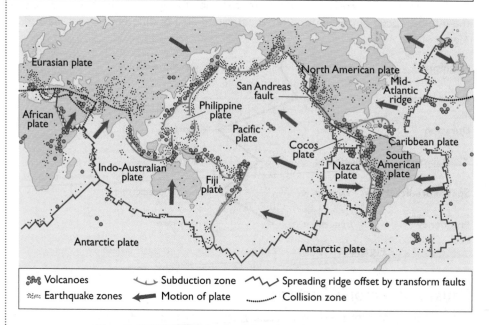

**Figure 4** Global distribution of earthquakes and volcanoes

**Knowledge check 3**

What are the origins of the word 'tsunami'?

Tsunamis ('harbour waves') are waves caused by the rapid movement or deformation of the seabed. This motion can be initiated by undersea earthquakes, volcanoes, landslides or slumps. They are classified as secondary hazards since they are events caused by a tectonic event.

The **distribution** of earthquakes and volcanoes (Table 2, Figures 4 and 5) has a distinct geography. There are two important areas of activity: the oceanic fracture zone (OFZ) and the continental fracture zone (CFZ).

- The **OFZ** is a belt of activity through the oceans along the mid-oceanic ridges, coming ashore at Africa, the Red Sea, the Dead Sea Rift and California.
- The **CFZ** is a more important belt of activity (passing through more populous areas). It flows through the mountain ranges of Spain, via the Alps to the Middle East, to the Himalayas and the East Indies and then circumscribes the Pacific.

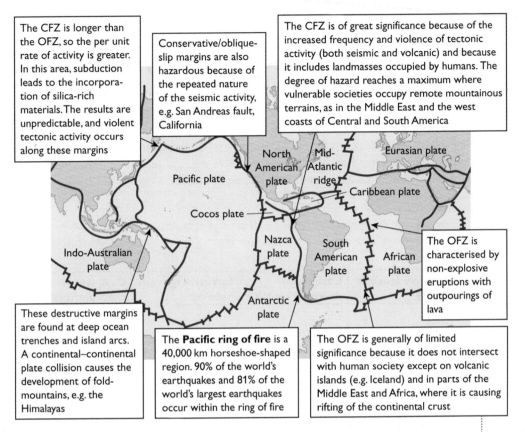

The CFZ is longer than the OFZ, so the per unit rate of activity is greater. In this area, subduction leads to the incorporation of silica-rich materials. The results are unpredictable, and violent tectonic activity occurs along these margins

Conservative/oblique-slip margins are also hazardous because of the repeated nature of the seismic activity, e.g. San Andreas fault, California

The CFZ is of great significance because of the increased frequency and violence of tectonic activity (both seismic and volcanic) and because it includes landmasses occupied by humans. The degree of hazard reaches a maximum where vulnerable societies occupy remote mountainous terrains, as in the Middle East and the west coasts of Central and South America

North American plate

Mid-Atlantic ridge

Eurasian plate

Pacific plate

Caribbean plate

Cocos plate

Nazca plate

South American plate

African plate

The OFZ is characterised by non-explosive eruptions with outpourings of lava

Indo-Australian plate

Antarctic plate

These destructive margins are found at deep ocean trenches and island arcs. A continental–continental plate collision causes the development of fold-mountains, e.g. the Himalayas

The **Pacific ring of fire** is a 40,000 km horseshoe-shaped region. 90% of the world's earthquakes and 81% of the world's largest earthquakes occur within the ring of fire

The OFZ is generally of limited significance because it does not intersect with human society except on volcanic islands (e.g. Iceland) and in parts of the Middle East and Africa, where it is causing rifting of the continental crust

**Figure 5** The geography of tectonic activity

Earthquakes and their associated volcanic activity at tectonic boundaries can sometimes be likened to a giant 'zip' where energy is gradually transferred along the plate boundary. The seismic 'unzipping' at this boundary can create a series of interrelated tectonic events. For example, the tectonic activity from the Tohoku (Japan) 2011 earthquake was described as a 'megaquake' along the subduction zone. The energy from this event 'rippled around the world' (*New Scientist*, April 2011). Bursts of energy on March 11 ripped through four separate patches that all individually generated quakes over an area of about 1000,000 km$^2$. There was also an unprecedented slip of 60 m.

There are two different types of crust: continental and oceanic (Table 3).

**Table 3** The Earth's crust

|  | Oceanic | Continental |
|---|---|---|
| Depth (km) | 5–9 | 30–70 |
| Density (g cm$^{-3}$) | 3.0 | 2.7 |
| Main rock type | Basalt | Granite |

**Examiner tip**

Make sure that you have good scientific knowledge of earthquakes and volcanoes at different types of plate boundary, i.e. technical information and vocabulary.

**Research activity**

How are the characteristics indicated in Table 3 important in determining the nature and impacts of both earthquake and volcanic activity? Key reference: USGS website: **http://vulcan.wr.usgs.gov/Servers/earth_servers.html**

# Enquiry question 2

## What impact does tectonic activity have on landscapes and why does this impact vary?

**Knowledge check 4**

Know the characteristics of the different types of lava/magma.

Both earthquakes and volcanoes can have significant impacts on landscapes, and create specialist features or morphology. Volcanoes, for example, can create impressive volcanic cones (shield, acid cone) and lava plateaus (e.g. Deccan plateau in India). Generally less imposing are the intrusive volcanic features such as batholiths, sills and dykes, often associated with oozing of magma.

**Research activity**

Copy and complete the table to consider differences in landscape impacts between intrusive and extrusive igneous activity. Support each type with a located example, including facts and figures.

| Extrusive igneous activity | Case study/example/location | Intrusive igneous activity | Case study/example/location |
|---|---|---|---|
| | | | |

**Research activity**

Find out about divergent and convergent plate boundaries, normal and reverse faults, and strike-slip faults (right and left). Also research rift valleys at divergent boundaries, using the following websites as a starting point: **http://tinyurl.com/dlclqp; http://scign.jpl.nasa.gov/learn/plate4.htm**

The impacts of earthquakes on landscapes are often less dramatic and diverse. At divergent plate margins (where two oceanic plates are diverging, such as the mid-Atlantic), **rift valleys** may form. They can also develop where an area of continental crust is being stretched by divergence, such as in eastern Africa. In a **normal fault** (again, common at divergent plate margins) a cliff-like feature known as a fault scarp may form. These can range from a few metres to hundreds of metres in height and their length may continue for 300 or more kilometres.

# Enquiry question 3

## What impacts do tectonic hazards have on people and how do these impacts vary?

It is predicted that by 2025, 600 million people will be living in tectonically active areas. There are a number of reasons why people may live in tectonically dangerous or risky locations:

- People may stay in hazardous locations because of a lack of choice or alternatives. Very poor people may have to live in unsafe locations, close to volcanoes or in areas susceptible to earthquakes.
- Many people subconsciously weigh up benefits and costs, e.g. the economic benefits may outweigh the perceived risks of living on fertile farming land on the flanks of a volcano.
- Perceptions of hazard risks can be optimistic — having faith in technology to overcome/negate risks.
- There may be traditional, historical or cultural reasons for living close to volcanoes, e.g. people believing volcanoes were 'sleeping gods' — inertia has prevented people from moving away from such locations.

**Knowledge check 5**

For comparison, what are the approximate numbers of deaths from other disasters in the last 10 years?

The number of deaths from volcanoes in recent years runs at about 1,000 per year, which is far greater than the number of deaths for previous centuries. However, the number of deaths from this type of tectonic hazard is much lower than that for earthquakes or hydro-meteorological events such as floods and droughts.

## Hazard types

There is a range of hazard types associated with tectonic events:
- earthquake, e.g. ground displacement, landslides, liquefaction, tsunamis, fires
- volcano, e.g. explosive blast, lava flows, ash flows and ash falls, mud flows, release of poisonous gases

Some areas show concentrations of tectonic and other hazards. These are known as **multiple hazard zones**. For example, the Philippines in southeast Asia is at risk from a number of hazards, including tectonic and hydro-meteorological hazards (see Figure 7).

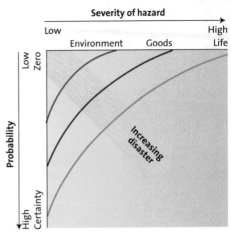

**Figure 6** Relationship between severity of hazard, its probability and the degree of risk

**Research activity**

Research the above hazard types (and other resulting hazards/ secondary hazards, e.g. glacial outbursts caused by a volcano under a mass of ice) to develop a selection of case studies that can be used to contextualise within the exam.

Key references for plate boundaries: **www. platetectonics. com/book/page_5. asp**; NASA Earth Observatory: **http:// tinyurl.com/3z8hkzv**

**Typhoons (hydro-meteorological)**
- A significant hazard (4–9 per year)
- Average annual death toll 529; cost £90 million
- Agricultural economy is the worst affected

**Earthquakes (tectonic)**
- Islands in zone between Pacific and Eurasian plates, with trench and subduction zones to the west
- In last 400 years, there have been about 65 destructive events
- The worst, in 1990 near Rizal City, caused 1,700 deaths and cost £300 million

**Floods (hydro-meteorological)**
- Flooding common, particularly as a result of high-intensity monsoon and typhoon rain
- Deforestation has increased flood magnitude and impact

**Tsunamis (secondary tectonic)**
- Coastal areas at risk from locally generated tsunamis and others from Pacific ring of fire
- Local tsunamis have an arrival time of 3–5 minutes
- From 1600 to 1980, 27 events recorded
- The worst (1976) killed 5,000

**Volcanoes (tectonic)**
- 200 volcanoes in the Philippines, 17 active
- Most famous is Mt Pinatubo, which erupted in 1991, causing 350 deaths; 58,000 people were evacuated
- Lahars had significant impact, especially in river basins close to volcano

**Figure 7** Hazards in the Philippines

## Impact types

The impacts that tectonic hazards have on people can be grouped into social and economic (Table 4).

**Table 4** Impacts of tectonic hazards

| | Volcanoes and earthquakes — example impacts |
|---|---|
| Social | Human costs, e.g. primary, secondary and tertiary casualties. Hazards can also cause misery and suffering as well as poor health, e.g. loss of electricity/services/infrastructure |
| Economic | Cost of repairing damage. Indirect costs/secondary impacts, such as loss of earnings, loss of tourism income |

Note that tectonic hazards can have *beneficial impacts* on people. These include geothermal sources of energy, rich fertile soils and minerals such as gold. There are also potential income streams associated with tourism foreign exchange, e.g. visitors to Iceland.

Since 1950 the number of reported disasters for all hazard types (hydro-meteorological, biological and geological) shows an average rising trend. But for tectonic disasters (earthquakes, tsunamis and volcanoes) there is generally a fluctuating trend that does not show a noticeable decrease or increase. Some researchers have provided data to show a slight increase in volcanic activity. This remains controversial.

## Trends in impact

Impacts can be measured in a number of ways, but typically they are reported in terms of number of deaths or casualties together with economic impact.

Figure 8 shows the number of deaths from geological hazards (including landslides/mass movements). Impacts seem to have been relatively constant over time. Note that there are fewer deaths and casualties associated with tectonic hazards compared to other hazard types such as floods and droughts.

**Figure 8** Deaths from geological hazards

However, as a researcher, you should question the reliability of data — who provided them and how were they collected? Deaths from earthquakes and volcanoes in remote areas are typically underreported. Impacts need to be studied at different scales and in countries with different states of development.

**Knowledge check 6**

What is a 'trend'?

**Examiner tip**

Keep up to date with recent examples of hazards and their associated impacts, especially for 'mega-events' where the casualties/death tolls may not be known for several months following the event.

**Research activity**

What evidence is there from the data that trends in impact have been influenced by anomalous years? Research additional data and statistics from the CRED website: **www.emdat.be/Publications/publications.html**

# Enquiry question 4

# How do people cope with tectonic hazards and what are the issues for the future?

Figure 9 shows how responses to tectonic hazards vary according to a number of factors, both physical and human.

**Figure 9** Factors that influence human response to hazards

## Range of responses

The response to natural hazards can occur at a range of scales, from the individual within the local community to regional, national and international level. Various 'players' may be involved in hazard response (Table 5).

**Knowledge check 7**

What are the other terms for 'players'?

**Table 5** Some of the players involved in hazard response

| | |
|---|---|
| **National government** | National policies, civil protection and defence, public information |
| **Local government** | Operational polices at the local level, e.g. rescue, welfare, medical, transport, supply of emergency aid |
| **Scientists/academics/ educators** | Researching, understanding causes, producing hazard (risk) maps. Also dissemination of information to public — raising awareness |
| **Insurers** | Risk assessment prior to hazard. Finance and assistance post hazard |
| **Planners** | Reducing risk by land-use zonation and planning, e.g. restricting development on floodplain |
| **Relief agencies** | Post-disaster aid, e.g. International Red Cross following the 2010 Haiti disaster |
| **Engineers and architects** | Design of buildings and infrastructures, e.g. Japanese technologists |
| **Emergency practitioners and services** | Police, medical, fire, traffic control — may also include army. Coordination is an important role |
| **Media** | In highlighting the tectonic hazard event (magnitude, scale, location, impact, duration etc.). Take on a warning role, especially important for more remote communities |
| **Communities — bottom up** | Important role in managing situations, as well as in terms of community preparedness and education |

# Specific strategies — before, during and after the event

The choice of strategy will vary during the different phases of a hazard. Figure 10 is an attempt to model the impact of a disaster from pre- to post-disaster. It also considers the role of emergency relief and rehabilitation (see 'players' on p. 15).

(see 'players' on p. 15)

**Knowledge check 8**

What are the advantages of understanding the disaster–response curve (or Park's model)?

**Examiner tip**

It may be a good idea to memorise a model such as the disaster–response curve for use in the exam. Make sure you replicate it correctly, with clear annotations that are linked to the focus of the question set. Don't repeat yourself in the accompanying text.

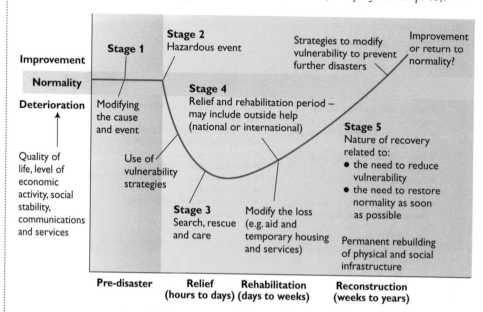

**Figure 10** A disaster–response curve

Figure 11 gives a useful framework for response analysis. Approaches to managing the tectonic hazards vary by time as well as place.

**Figure 11** Response-analysis framework

**Knowledge check 9**

What is the difference between risk and vulnerability?

There is also an interesting balance that needs consideration between increasing vulnerability and risk — see Table 6.

**Table 6** The balance between vulnerability and risk

| Increased vulnerability | Reduced vulnerability |
|---|---|
| • Population growth | • Warning and emergency-response systems |
| • Urbanisation and urban sprawl | • Economic wealth |
| • Environmental degradation | • Government disaster-assistance programmes |
| • Loss of community memory about hazards | • Insurance |
| • Ageing population | • Community initiatives |
| • Ageing infrastructure | • Scientific understanding |
| • Greater reliance on power, water, communication systems | • Hazard engineering |
| • Over-reliance on technological fix | |

### Research activity

Copy and complete the table below (earthquakes vs volcanoes) to give an analysis of types of responses to the management of tectonic hazards. You could also support each comment with a located example.

| | Advantage | Disadvantage |
|---|---|---|
| Land-use control/design resistant buildings | Can save lives; can be easily regulated through planning controls etc. … | |
| Education and training | | |
| GIS/remote sensing | | Cost (technology gap — MEDC vs LEDC); reliability; coverage… |

**Summary**

- You should research why some people suffer risk from earthquakes, others from volcanoes and some from both. How important are global plate boundaries (transform, convergent etc.) in determining the level of risk and exposure?
- What are hazard profiles and to what extent are they useful in developing knowledge and understanding of how tectonic hazards might vary?
- Explore how increasing urbanisation on a world scale (but especially increasing in emerging economies) is changing the vulnerability of certain populations.
- Some areas of the world suffer from tectonic-related disasters only occasionally, whereas others have a much higher frequency of disaster. Examine why this is, and the complex economic, social, environmental and political forces that have led to these imbalances.
- Investigate the role of level of development in determining impacts and possible risk, both between countries and within larger states, e.g. the USA and China.
- You should research examples and case studies of a range of ways of managing tectonic risk, including do nothing and abandonment. It is important to be able to evaluate the costs and benefits of individual approaches, also examining how and why approaches can vary.
- Investigate the global location of tectonic margins where risk is concentrated; this should include a consideration of the vulnerability of people. Are they suitable locations for world cities? Where are the other tectonically risky locations away from plate boundaries?
- Research how political stability, careful land management and new technologies might help reduce risk in tectonically active areas. What role does the local community have in making itself less vulnerable to potential disasters?
- Contrast different types of technology that could be used to both manage and predict tectonic hazards (earthquakes vs volcanoes) and how technology can be used to assist in disaster recovery, e.g. GIS and social networking (used in Haiti 2010, for instance).
- Consider the role of players such as IGOs and NGOs in managing tectonic hazard impacts and disasters, both in the short and longer term.
- Investigate whether or not longer-term trends in risk and exposure are really changing or if it is more to do with accuracy and quality of reporting (including the role of technology such as Twitter in communicating information).

# Cold environments: landscapes and change

## Introduction

Cold environments occur on a large percentage of the Earth's surface. There are also many other locations that show evidence of having been cold environments in the past as a result of the historically cooler climatic conditions that have existed in parts of the world through geological time.

These cold environments possess a variety of landform types and features, reflecting the diverse nature of the physical and human processes that have helped shape them. These processes are greatly influenced by current and past climatic conditions, especially by shorter- and longer-term variations in temperature.

Cold environments can provide many opportunities for human activity and as a result there can be conflicts in terms of management, especially exploitation vs protection. Different attitudes exist within the various interested parties and this can lead to significant conflicts between them.

You will need to research a range of glacial and periglacial environments, both present and relic. In particular, you will need to find out about the distribution of different types of cold environments.

**Examiner tip**

When you research case studies and smaller examples make notes on all the enquiry questions, with a log of sources in case you need to add details on one area once the research focus pre-release material is available.

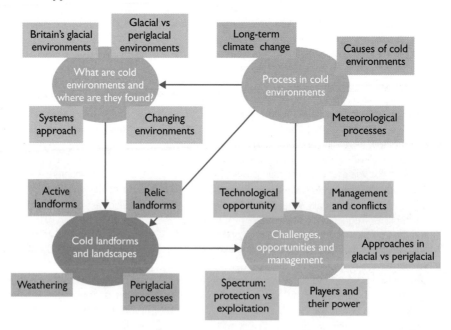

**Figure 12** Topic concept map

Figure 13 The synoptic context

Research activity

Use Table 7 as an initial stimulus, then carry out your own research audit to obtain background sources of information. Make a note of any key references used.

Table 7 Checklist for cold landscapes

| Enquiry question 1<br><br>What are cold environments and where are they found? | Enquiry question 2<br><br>What are the climatic processes that cause cold environments and what environmental conditions result from these? | Enquiry question 3<br><br>How do geomorphological processes produce distinctive landscapes and landforms in cold environments? | Enquiry question 4<br><br>What challenges and opportunities exist in cold environments and what management issues might result from their use? |
|---|---|---|---|
| • What are the different types of cold environments?<br>• Glacial vs periglacial<br>• Global patterns and distribution<br>• Short vs long-term changes in distribution<br>• Glacials vs interglacials (especially history in British Isles)<br>• Landscape systems<br>• What is mass balance? | • Short- vs long-term changes in climate — atmospheric 'triggers'<br>• Global systems as distributing heat energy<br>• Where were past glacial and periglacial environments?<br>• Climate change and future for cold environments | • How different are ice sheet and valley glacier landscapes?<br>• How different are upland and lowland ice landscapes?<br>• Scale and energy of impacts<br>• Relic vs active ice landscapes<br>• Periglacial landscapes and landforms | • What are the main challenges in living in/exploiting cold environments?<br>• What opportunities do cold environments present?<br>• How and why are there different attitudes to protection or exploitation of cold landscape resources? |
| Key refs used | Key refs used | Key refs used | Key refs used |

# Enquiry question 1

# What are cold environments and where are they found?

Cold environments are icy landscapes that occur in (a) polar areas and (b) high mountainous regions of the world. They are characterised by very cold temperatures together with lots of snow and ice, and few plants and animals. They have also experienced relatively little commercial exploitation and have low human population densities. But not all 'cold' environments are the same and they can be subdivided into: (a) glacial areas, (b) tundra and (c) alpine areas or high mountain ranges.

## Understanding the basic concepts

Many cold environments can be understood by way of a systems approach, i.e. by thinking about three distinct elements: *inputs*, *processes/transfers* and *outputs* (Figure 14).

**Knowledge check 10**

Give single-sentence definitions for: inputs, processes and outputs.

**Figure 14** Inputs, processes and outputs

The balance between the inputs and outputs of snow and ice in a glacier (Figure 15) is called the **mass balance**. This can be either positive or negative (Figure 16).

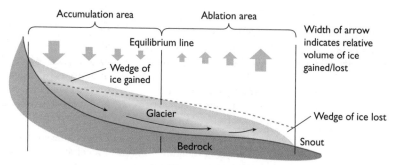

**Figure 15** The glacial system

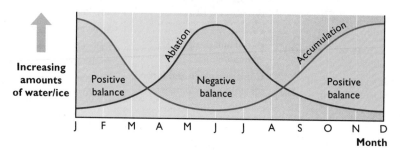

**Figure 16** The net budget in a northern hemisphere glacier

**Knowledge check 11**

What is the firn line?

There will often be seasonal variations in the budget with accumulation exceeding ablation in the 'winter' and vice versa in the 'summer'. It is possible that there will be an advance during the year even if the net budget is negative and some retreat even when it is positive. Even in retreat, a glacier may advance under gravity in front of the **firn line**, so it can appear that a retreating glacier is actually advancing.

Some other important terminology is summarised in Table 8.

**Table 8** Glacial terminology

| Frequency and magnitude | Frequency refers to how often an event happens or repeats itself. Magnitude refers to the size of an event. In glacial systems high frequency/low magnitude events include the daily abrasion carried out beneath a glacier. Some events happen very infrequently (low frequency) but can be catastrophic in terms of power and impact on the landscape, e.g. a glacier burst (jökulhaulp). |
|---|---|
| Equifinality | The notion of equifinality states that in an open system the same end result can be reached by many different means. In glacial environments this means that different processes may produce the same landform or landscape feature. For instance, outwash plains may be produced by a small number of major glacial bursts, as well as by small, incremental amounts of fluvioglacial deposition over a longer period of time. |

| Dynamic equilibrium | Natural systems have a tendency towards achieving a steady balance. If something happens to disturb the equilibrium of a balanced system, the system itself produces a dynamic response in order that equilibrium can be achieved. But equilibriums can change over time and under different conditions. For example, in glaciers an increase in accumulation leads to a growth in the glacier. This in turn advances the equilibrium/ firn line until the system returns to a stable state. |
|---|---|

## Changing cold environments

The distribution of cold environments has changed throughout geological and more recent history. In the past, cold landscapes have been more extensive than at present (Table 9) (nearly all the ice on Earth today is locked up in Greenland and Antarctica). One controversial theory, 'Snowball Earth' suggests that at one point in history the whole planet was covered in ice. Even if this contentious idea is not true, what is certain is that there have been repeated glacial periods (glacials) separated by warmer interglacial times (Figure 17). The last glacial maximum occurred about 20,000 years ago — this was the latter part of the Pleistocene ice age.

**Table 9** Ice coverage in 1970 compared with maximum for selected world regions

| Region | Area of ice in 1970 (km²) | Maximum area during glacial maximum (km²) — estimated |
|---|---|---|
| Antarctica | 12.5 million | 13.8 million |
| Greenland | 1.7 million | 2.3 million |
| Asia | 115,000 | 3.9 million |
| South America | 26,000 | 870,000 |
| Europe | 3,600 | 37,000 |

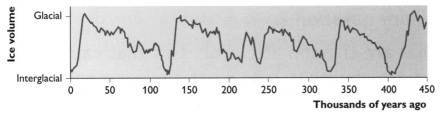

**Figure 17** Glacial and interglacial periods of the last 0.5 million years

Oxygen-isotope ratios can be used to reconstruct past climates. Note the fluctuations during the Devensian period in Figure 18 — a series of glacials and interglacials.

**Figure 18** Oxygen-isotope ratios

**Knowledge check 12**

What are the other techniques that can be used to interpret past climates?

## Past ice in Britain

In the past, when the climate was much colder, there were several significant glacial environments in Britain. They were found mainly in Scotland, the Lake District and North Wales as well as in parts of Ireland. Britain experienced several glacial periods during the last 2 million years of the Quaternary period (Figure 19).

**Knowledge check 13**

What is meant by the term 'glacial maximum'?

**Research activity**

Find out more about the ice age in one part of Britain or overseas. Key reference for impact and evidence in north Norfolk: **http://tinyurl.com/3nue58w**

**Figure 19** The glacial maximum in Britain

# Enquiry question 2

# What are the climatic processes that cause cold environments and what environmental conditions result from these?

## Factors affecting temperature

The low temperatures that characterise cold climate environments are the result of four factors:

**Knowledge check 14**

What are absorption, scattering and reflection?

1 **Latitude** — in high latitude areas (i.e. 60°–90°N and S of the equator) the sun in the sky is low. Incoming solar radiation passes through a greater thickness of atmosphere compared to mid-latitude locations (see Figure 20). Energy is lost due to absorption, scattering and reflection (e.g. from surfaces such as cloud tops).

Also, the low angle of incidence means that the remaining solar energy is spread over a wider area, and much of this incident energy is reflected back into the atmosphere from the surface of snow and ice (high albedo or reflectivity).

The other implication of latitude is the resultant pressure systems. Cold, dense air tends to sink at or near the poles and produces high-pressure polar anticyclones. Air from these pressure systems moves slowly outwards, spreading into the high mid-latitude areas causing low temperatures.

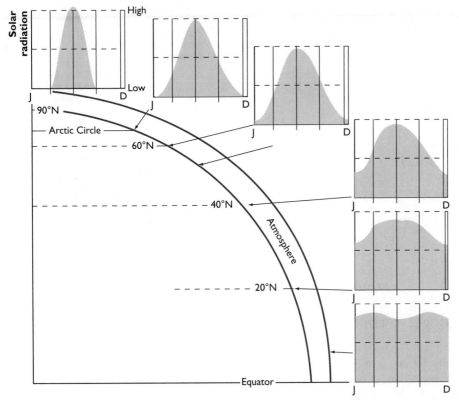

**Figure 20** The effect of latitude on incoming solar radiation

2 **Altitude** — temperatures decline with altitude because air becomes thinner with height above sea level and is less able to absorb heat from outgoing, long-wave terrestrial radiation. The decrease in temperature with altitude is called the environmental lapse rate and on average is about 6°C per 1,000 m. This fall in temperature explains why glaciers can exist near the equator, but only at altitudes above 5,000 m, e.g. Cotopaxi and Cayambe volcanoes in Ecuador.

3 **Continentality/distribution of land and sea** — the ability of land and sea to absorb heat and radiate it back into the air above varies greatly. In general, land (soil and rock) warms up more quickly, and is able to release the stored heat at a faster rate compared to water. This means continental interiors are warmer than coastal areas in the summer, but much colder than coastal areas during winter because the land has lost its stored heat, whereas the sea is still radiating heat that it absorbed during the previous summer.

4 **Ocean currents** — The importance of ocean circulation has only been revealed relatively recently. It has now been demonstrated how the warm North Atlantic Drift (NAD), for example, partly controls temperatures in Western Europe. The current transfers heat energy from the Gulf of Mexico northwards across the Atlantic (Figure 21).

There are concerns that large influxes of fresh water (e.g. from the melting of ice sheets or greater discharge of rivers) can actually turn off this heat conveyor belt. A loss of the NAD would have significant impacts on Western Europe both in terms of environments and economies.

**Knowledge check 15**

Why are ocean flows in the North Atlantic especially important for global temperature regulation?

**Examiner tip**

Use the correct terminology to describe climate change vs global warming. The ideas and concepts are not interchangeable.

**Research activity**

Find out more about the longer-term causes of climate change. Which are the most significant? See the British Antarctic Survey for more details: **www.antarctica. ac.uk/bas_research/ science/climate/ climate_change.php**

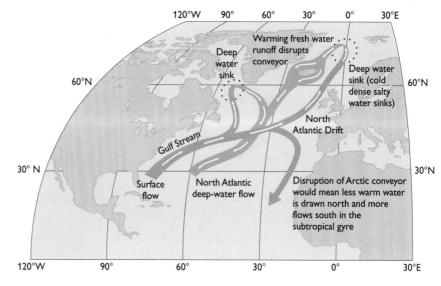

**Figure 21** The path of the North Atlantic Drift oceanic current

## Longer-term climate change

Throughout the Earth's history temperatures on the surface of the planet have varied considerably (see Figure 17). Many different causes of these medium and longer-term climatic changes have been proposed and include changes in the Earth's orbit, the tilt of the Earth's axis, solar radiation, position of the poles, the amount of water vapour in the atmosphere, distribution of land and sea, flows of ocean currents and the amount of carbon dioxide in the atmosphere.

# Enquiry question 3

# How do geomorphological processes produce distinctive landscapes and landforms in cold environments?

Geomorphology is the study of landforms, landscapes and the processes that create them. Areas that have been glaciated usually have distinctive landscapes because ice has significant power to modify existing landforms. A well-used example is where ice in a valley glacier changes the scale and shape of a pre-existing river valley (from v-shaped to more of a trough shape) — see Figure 23.

## Processes of erosion

**Examiner tip**

Good technical knowledge of these processes and precise details of their operation, for example in producing characteristic landforms, is important for many responses. The same is true for sub-aerial processes.

Two main erosional processes, abrasion and plucking, are responsible for some of the most impressive erosional landforms on the planet:

● **Abrasion** occurs where glacier ice is in contact with the valley bottom and sides. Ice at the bottom of the glacier carries debris and scratches or abrades the bedrock surface. Figure 22 shows the factors that can affect abrasion. The process of abrasion is evident in the form of striations — scratches on bedrock that indicate the direction of ice flow.

**Figure 22** The factors that can affect the amount and rate of glacial abrasion

- **Plucking**, or 'joint block removal' is also dependent on the movement of ice. Ice freezes to jointed bedrock and then the rock becomes incorporated into the sole of the glacier as it moves. This is only possible due to basal pressure melting whereby the base of the glacier can thaw and re-freeze (re-gelation) due to changes in the pressure exerted by the overlying ice.

Rates of glacial erosion will vary enormously in both space and time. In 1968, two researchers calculated that mean annual erosion for active glaciers is between 1,000 and 5,000 m³. In comparison, ice 100 m thick flowing at 250 m per year in the French Alps managed to lower a marble plate by 36 mm in one year.

**Research activity**

Find out more about how the factors shown in Figure 22 will affect the amount and rate of abrasion carried out by a glacier. Also, research examples of different glaciers to illustrate these various factors. Visit the World Glacier Inventory: **http://nsidc.org/data/g01130.html**

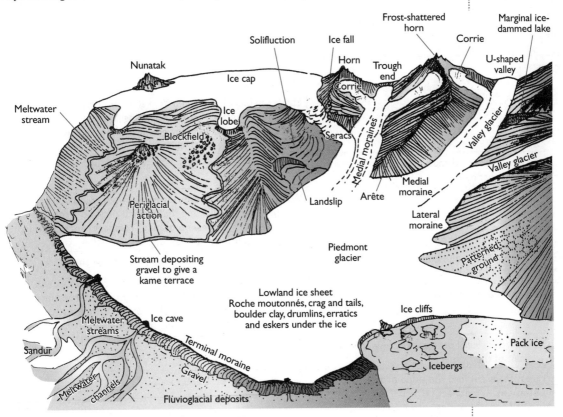

**Figure 23** Range of glacial landform types

Sub-aerial processes may improve the efficiency and impact of abrasion and plucking by providing additional weathered rock debris:

- **Freeze–thaw** — the pressure exerted by water changing state and expanding into ice. This process, when repeated, can be highly effective in the destruction of heavily jointed rock, particularly material exposed above the level of ice.
- **Dilatation** — rocks can fracture when overlying pressure is released. The ice melts, losing weight and therefore exerting less downward pressure. The rocks of the valley floor/sides expand and fracture as the weight is removed.

## Distinctive landforms

An important task for this module is to research both upland and lowland landforms of glaciation. (Figure 23 on p. 25 shows an example of the range of landform types.) For each landscape feature (erosional, depositional, periglacial, upland, lowland etc.) you will need to know about:

- processes which formed the feature (including the role of weathering)
- scale and distribution (within the glacial system and regionally), plus characteristic shape/patterns
- examples: location, size, special characteristics etc.

You will also need to research **periglacial environments** — both above ground and below ground processes and features. Table 10 provides a list of periglacial features and locations that can act as a starting point.

**Table 10** Periglacial environments and example locations

| Feature | Examples |
|---|---|
| Ice wedge casts | Giant ice wedge casts at Stanton Harcourt and Baston in Lincolnshire |
| Patterned ground | Stripes in Lincolnshire and Yorkshire wolds; nets and stripes on Thetford Heath, East Anglia; centre of nets correspond to chalk-loving grasses, while the 'net' reflects heathers growing on thick sands |
| Cryoturbation structures (deformation of sediments near surface, for example frost heave) | Isle of Thanet, Kent; East Anglian chalklands |
| Pingos | Walton Common, Cambridgeshire, where 'scars' are superimposed, indicating that several cycles of ground-ice growth are represented; many rampart repressions in valleys of southwest Wales, e.g. Cletwr Valley |
| Thermokarst scenery-slumps and depressions caused by melting ground-ice | Walton Common, Grunty Fenn, Cambridgeshire |
| Head deposits formed through frost shattering, solifluction and sliding | Sarsen stones — streams of surface boulders running along the valley bottom of the Marlborough Downs, e.g. Coombe rock, consisting of chalk, flint and mud |
| Asymmetric valleys (periglacial processes acting unequally on slopes as a result of their aspect) | Chilterns |
| Nivation terraces (steps cut into the slope profile caused by prolonged nivation) — frost-action, mass wasting and meltwater erosion beneath melting snowdrifts | Slopes of Cox Tor, Dartmoor |
| Tors | Pennine tors (gritstone); Stiperstones, Shropshire (quartzite); Dartmoor tors (granite) |

**Examiner tip**

Produce an annotated A3 map of the UK, which shows the locations of periglacial features, outlining their processes of formation. This could potentially be adapted and simplified for replication in the exam.

**Knowledge check 16**

Define 'periglacial' and 'weathering'.

# Enquiry question 4

## What challenges and opportunities exist in cold environments and what management issues might result from their use?

Cold landscapes offer a range of threats, challenges and opportunities to people. Although only a small proportion of the world's population live in cold environments, many of us are either directly or indirectly influenced by cold locations. Periglacial and permafrost environments present their own challenges and opportunities.

**Research activity**

Find out more about the wide range of challenges (these can include hazards) that exist in cold environments. Put your findings into a table, similar to the example provided.

| Threat | Description | Located example |
|---|---|---|
| Ice avalanche | Ice 'calved' from the snout in steep areas of relief. Sometimes ice avalanches occur together with landslides and snow avalanches... | Mt Iliamna, Alaska, in 1980 produced 20 m³ of ice.<br>Speeds of up to 100 km h⁻¹... |
| Flood | Catastrophic outburst floods (jökulhlaups). Meltwater is suddenly released... | ... |
| Mud flow and lahar | ... | ... |

Other challenges may be less obvious, but still difficult for people who live in such environments, e.g. problems of low agricultural productivity and difficulty in establishing and maintaining transport infrastructure.

**Research activity**

Over a period of 4–6 weeks, scan the international newspapers and their corresponding websites to identify any news items linked to cold environment challenges, e.g. management of fragile ecosystems, sea level rise (melting ice NOT thermal expansion), avalanches, floods, etc. Make brief notes on each so that they can be used as case studies in preparation for the exam. Key reference: many newspapers of the world can be accessed at **www.newseum. org/todaysfrontpages/flash/**

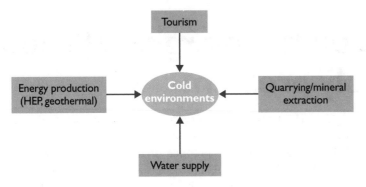

**Figure 24** Cold environments also present many opportunities

**Research activity**

Investigate how technology can be used in both periglacial and glacial environments to overcome challenges linked to one element of exploitation, e.g. building a new HEP station, maintenance of fragile ecosystems.

Ensure that your facts, figures and data are up to date. Think about the advantages/disadvantages of this type of tourism, especially in the context of climate change. Glenshee Ski Centre attracted 171% more skiers (up from 46,355 in 2009 to 125,831 in 2010).

Since the mid-nineteenth century, glaciers have provided the basis for particular types of tourism and now form the cornerstone of many local economies. For example, the tourist industries of Greenland and Iceland are based largely on their spectacular glacial landscapes. Some National Parks are based on present day glaciers, e.g. Glacier National Park in Montana, or on relic glacial landscapes, e.g. the Lake District in the UK.

**Summary**

- You should research why some places experience different types of cold environment: polar (high latitude) and alpine (high altitude); recognise the current distribution of ice and cold lands.
- Be able to contrast different types of glacial system, i.e. glacial vs periglacial vs fluvio-glacial, and recognise the importance of systems theory in understanding change.
- Explore how and why this pattern of ice and cold conditions has changed over time, and what have been the drivers (both internal and external 'forcings').
- Recognise the complexity of the Earth's climate system and triggers and tipping points, and how these can have significant impacts on the amount of global ice and albedo.
- Explore and understand the link between levels of insolation and the precise location of cold environments. What other factors may influence these patterns of distribution?
- Investigate how the climate (and weather) in Britain has changed considerably over the Pleistocene and especially the Holocene (last 10,000 years).
- Appreciate the erosional and sub-aerial processes that are operating in cold environments; how different landforms are a result of such processes and other factors, e.g. geology.
- Investigate how periglaciation involves a different set of processes and will create a unique set of specialist landforms.
- Investigate the possible climate change future scenarios — predictions in areas such as the Arctic as well as influencing extreme precipitation events.
- How might the impacts of global warming present additional challenges and opportunities for certain people living both within and outside cold environments?
- Investigate how new technology may reduce the barriers to exploiting and operating in extreme cold environments.
- Consider the role of players such as IGOs and NGOs in managing cold environments and how this can create a complex set of challenges for different groups who may have competing needs.
- Investigate whether or not cold environments really can be sustainably managed — is this simply protectionism or can it be managed in other ways?

# Life on the margins: the food supply problem

There is enough food in the world for everyone; the problem is its uneven distribution and unequal access. Over 50% of the world's population live in low-income, food-deficit countries that are unable to produce or import enough food to feed their people. Most of the world's hungry people are in the developing world, but 19 million lived in the developed world in 2010. Life is on the margin of survival for these people in both rural and urban areas. Over one third of all children are malnourished and

6 million children a year die of causes related to malnutrition. One leading player in the food supply problem, the UN Food and Agriculture Organization (FAO), organises World Food Day on 16 October each year, to highlight its aim of food security: 'when all people have physical and economic access at all times to enough nutritious, safe food to lead healthy and active lives.'

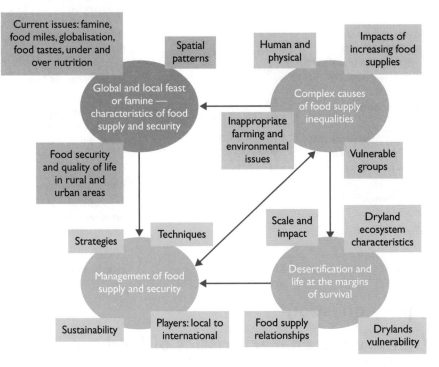

**Figure 25** Topic concept map

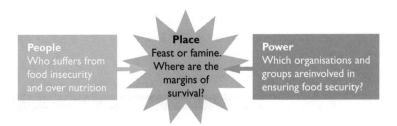

**Figure 26** The synoptic context

**Research activity**

While thinking of the synoptic aspects of the course, a simple checklist based on the topic/ enquiry questions from the full specification will help you separate your research into manageable chunks (see Table 11). Key references: overviews, case studies, quizzes: **http://anthro.palomar.edu/change/default.htm**; 2008 World Food Summit: **http://news.bbc.co.uk/1/hi/in_depth/7432583.stm** and **www.fao.org/newsroom/ en/focus/2008/1000829/index.html**; 'The silent tsunami', Food crisis of 2008: **www.mssrf.org/fs/index.htm** and **http://tinyurl.com/4y8zt4**

**Table 11** Checklist for life on the margins

| Enquiry question 1 | Enquiry question 2 | Enquiry question 3 | Enquiry question 4 |
|---|---|---|---|
| **What are the characteristics of food supply and security?** | **What has caused global inequalities in food supply and security?** | **What is the role of desertification in threatening life at the margins?** | **How effective can management strategies be in sustaining life at the margins?** |
| • What type of socioeconomic current issue?<br>• Is it a global or more localised issue?<br>• Any environmental issues?<br>• Why does food supply vary spatially?<br>• How does food security affect quality of life? | • Classify the causes into economic, social, environmental plus direct and indirect: long and short term<br>• Is population pressure involved?<br>• Is type of food production technique involved, and effects on environment?<br>• Identify which groups of people most affected by food insecurity<br>• Why are they affected? | • Scale of problem<br>• Physical climatic causes?<br>• Human causes?<br>• Role of increased food supply<br>• Is salinisation a problem?<br>• Why are dryland ecosystems so vulnerable?<br>• How is food supply affected by desertification? | • Who is involved in management?<br>• Role of organisations/ groups involved from local to global<br>• Types of strategies/ techniques used (e.g. trade, aid, increased food production by high and low tech)<br>• Reasons for those<br>• How effective are the organisations/groups and strategies/techniques?<br>• Role of sustainability |
| Key refs used | Key refs used | Key refs used | Key refs used |

# Enquiry question 1

# What are the characteristics of food supply and security?

## Current issues in 'Life on the margins'

- Food insecurity — the unreliability of quality food supply can lead to famine and ultimately death (see Figure 27).

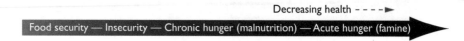

Figure 27 Food insecurity

- The areas most at risk are often arid and semi-arid areas such as the Sahel, but vulnerable groups are growing as the development gap widens, from Mumbai to New York, Bolivia to Cambodia — with a concentration in sub-Saharan Africa (see Figure 29).
- Food transition issues: overnutrition and obesity. This involves the globalisation of food tastes and the increased consumption of protein, fats and sugars by rising middle classes especially in urban areas. Globally, obesity has now reached epidemic proportions.
- Food suppliers: the rise in power and influence of mega-supermarket chains and TNCs plays a large role in undercutting local food supply chains and producing cheap fast-food alternatives, especially for urban populations.

**Knowledge check 19**

Name some TNCs that are directly related to the food security debate.

Edexcel A2 Geography

- Environmental issues: the negative effects of the scale and intensity of food production. Since the 1960s, issues have included the Green Revolution, GM crops, deforestation and growing biofuels in place of staple food crops.
- Food miles and footprints: the distance food travels from producer to consumer and environmental impacts of food supply.
- Food supply and prices crises in 2008 and 2010–11: causing global food scarcity.

## What is food security?

Figure 28 shows that food must meet physiological requirements in terms of quantity, quality and safety and must be socially and culturally acceptable. Food security is determined at local and national scales by stability in availability, access, and use/utilisation. Where the nutritional status is low, there is no food security even though there might be a surplus of food supply at the national or regional level. It is clear that it will be difficult for managers to solve problems of insecurity, which often have complex causes.

There is a difference between:
- Transitory food insecurity — which may be either cyclical/seasonal (e.g. gaps in staple food production like rice during a 'lean season') or temporary (from unpredictable climate shocks like drought and floods).
- Chronic food insecurity — where vulnerable groups are permanently unable to ensure their food needs, usually linked to poverty.

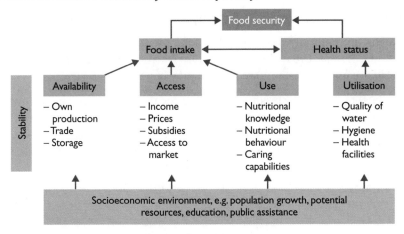

**Figure 28** Food security

To be food secure means that:
- Food is available — the amount and quality of food available globally, nationally and locally can be affected temporarily or for long periods by many factors including climate, disasters, war, civil unrest, population size and growth, agricultural practices, environment, social status and trade.
- Food is affordable — when there is a shortage, food prices increase, and while richer people will likely still be able to feed themselves, poorer people may have difficulty obtaining sufficient safe and nutritious food without assistance.
- Food is utilised — at the household level, sufficient and varied food needs to be prepared safely so that people can grow and develop normally, meet their energy needs and avoid disease.

**Examiner tip**
Use the BBC Business website to keep a close eye on commodity food prices (wheat, rice, sugar), which tend to be volatile. Price rises hit the developing world's poor hardest as they spend a high proportion of total income on food.

**Knowledge check 20**

What is meant by 'food security'?

The physiological effects of food insecurity include:

- Intergenerational problems — low calorie and nutrient intake slows metabolism, reducing ability to work and makes food more difficult to utilise. A hungry mother will have underweight babies who will have stunted growth, frequent illness, learning disabilities and reduced resistance to disease.
- Contaminated food and water can also cause illness, nutrient loss and often death in children.
- Acute malnutrition is the result of sudden weight loss due to starvation and disease, often leading to rapid death as it increases the risk of infection and vital organs stop working. It is treatable but it may have long-term effects on physical and mental growth.
- 'Wasting' means children are far thinner for their height than healthy children.

## Issues resulting from increased food production methods

Problems resulting from increased food production of common UK food types (e.g. orange juice, cooking oil, milk, chicken, beef, fish, sugar, fruit and vegetables) include:

- pollution issues, e.g. waste, chemicals, emissions
- loss of biodiversity
- rising ecofootprint
- food miles
- ethical issues, e.g. animal welfare, social issues (exploitation of workers)

**Knowledge check 21**

What is meant by 'food miles'?

## Why food supply varies spatially

Some areas are naturally more suited to food production, e.g. the great plains of America and Russia. Urban fringes have also been traditionally important in food supply, but are under constant threat from sprawling urban growth.

Physical factors limit food production unless technology is available to overcome temperature, water and nutrient deficiencies. Irrigation, chemicals and greenhouses are, however, costly. Agricultural advances in yields have shown there is a law of diminishing returns that limits even GM products as a supply source. Table 12 lists human and physical reasons for variation in food supply.

**Table 12** Factors that influence food supply variations from place to place

| Human factors | Physical factors |
|---|---|
| Accessibility of markets | Soil — nutrient store |
| Land ownership systems — rented land or that without secure tenure causes insecurity | Climate — seasonal changes |
| Inheritance laws — may be gender biased | Precipitation — amount, frequency, type |
| Market and trade patterns and regulations skewed in favour of more developed economies | Length of thermal growing season |
| Competition, which can be healthy but often unfair if subsidies, quotas, etc. involved | Relief — steep or waterlogged areas less useful |
| Government action and support | Aspect — slope angle |
| Role of businesses and TNCs, which now dominate research globally and are governed by profit margins, not necessarily food security of the poorest | Altitude — affecting temperature, water supply |
| Role of aid agencies in both long- and short-term food supplies | Hazards — tectonic, hydro-meteorological and biological. Northern China produces 58% of the country's food crops; in 2011 it suffered its worst drought in 50 years highlighting the increasing concern over climate change reducing agricultural production |

Contrast and compare two key case studies on food intensification, security and resulting socioeconomic and environmental problems: the EU's Common Agricultural Policy and the Green Revolution based in Asia and Latin America. Create a timeline for each to show changes, and why the new Green Revolution of the twenty-first century, targeting Africa, may make food security more sustainable. Key references: short podcasts of past and present Green Revolutions, food transitions and impacts on environment: **http://news. bbc.co.uk/1/hi/programmes/documentary_archive/6500041.stm**; Greening of the Green Revolution: **www.nature.com/nature/journal/v396/n6708/full/396211a0. html**; Africa's food crisis: **http://news.bbc.co.uk/1/hi/in_depth/africa/2006/africa_ food_crisis/default.stm**

**Knowledge check 22**

Explain what is meant by the Green Revolution.

## Food security in urban and rural areas

Traditionally food insecurity is a characteristic of rural areas, especially in poorer economies. Figure 29 shows that sub-Saharan Africa has the highest rate of undernourishment.

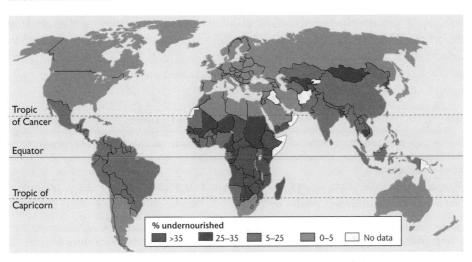

**Figure 29** Global distribution of undernourishment, 2003–05

India still ranks ninety-fourth in the UN World Food Programme Global Hunger Index of 119 countries. Large-scale famines have been controlled since independence in 1947, the Green Revolution has boosted food production, especially in the Punjab. However, according to the *Food Insecurity Atlas of India* produced by the Swaminathan Research Foundation, it now has the highest absolute number of undernourished people in the world, especially in small towns.

Urban food insecurity is not confined to developing nations: the New York City Coalition Against Hunger estimated that by 2006 1.3 million New York residents, including 400,000 children, lived in households with an insecure food supply. Michelle Obama made food security a headline in 2009 by helping in a Washington soup kitchen: food insecurity is on the increase globally. People on low incomes, wherever they are, will almost inevitably be more vulnerable to food security issues than those who are more affluent.

**Examiner tip**

Be careful not to assume that food insecurity is only a rural problem. Remember that over 50% of the world's population now lives in towns and cities.

# Enquiry question 2

# What has caused global inequalities in food supply and security?

## Food production and food security

Food production varies greatly. Between 1960 and 1990:

- World cereal production more than doubled.
- Food production increased by one third per head.
- Daily intake of calories increased by one third.
- Real food prices fell by almost half.

**Examiner tip**

Keep up to date with trends in food insecurity as changes are frequent. In 2009, over 1 billion people were food insecure but this dropped in 2010 largely as a result of food prices falling.

Yet in 2010, 925 million people, more than the populations of the USA, Canada and the EU, do not have enough to eat — 906 million of these are in developing countries. The FAO estimates that every 6 seconds a child dies, and the number of undernourished people in the world increased by 75 million in 2007, largely due to higher food prices.

Soil degradation, chronic water shortages, inappropriate agricultural policies and population growth threaten food production in many countries. While growing export crops such as coffee, cocoa and sugar produces export income, it can lead to a decrease in basic food production, causing hardship for people who are poor. These complex causes may be categorised into root and direct causes (Table 13).

**Table 13** Factors affecting food security

| | Direct causes | Root causes |
| --- | --- | --- |
| **Economic** | Income, poverty trap, land security/tenure, food supply from local, national imported production, aid, infrastructure (roads, storage, water), food hoarding | Trade restrictions, debt repayments |
| **Social** | Population growth, poor health and reduced labour (especially HIV), war: deliberate food destruction, gender equity | War and corruption, refugees and displacement, rise of middle class and changed food tastes |
| **Environmental** | Natural disasters of drought, desertification, floods, pests, overcropping and overgrazing, urban sprawl | Pollution and climate change especially drought |

Since 1992, the proportion of short- and long-term food crises that can be attributed to human causes has more than doubled, rising from 15 to over 35%.

**Research activity**

Fifty years ago, most counties in Africa were more than self-sufficient in food and yet now most are major importers and reliant on outside food aid. By 2008, the FAO warned that 27 sub-Saharan countries needed urgent help with food security. Gather evidence for Africa's systemic problems, based under the headings of:

- Economics (underinvestment, trade, aid, debt)
- Politics (government corruption and war)
- Health (HIV/AIDS)
- Demographic (natural increase and migrations)
- Environmental (hazards and resources)
- Key references: Famine Early Warning System: **www.fews.net/Pages/default.aspx**
- World Food Programme: **www.wfp.org**
- UN Environment Programme: **www.unep-wcmc.org/habitats/drylands/index.htm**
- International Food Policy Research Institute: **www.ifpri.org/**

# Population pressure and resource relationships

Academics are split between two opposing views on the relationship between population and resources, as summarised in Figure 30. However, both views may be appropriate at different scales.

- On a global level the growing suffering and famine in some developing countries may support a Malthusian scenario.
- On a national scale some governments have been motivated by increasing populations to develop their resources to meet growing demands, as in Mauritius. The Green and Gene Revolutions may also be used as evidence of 'technological fixes' to solve food insecurity especially famine. This idea underpins Boserup's theory.

**(a) Malthusian model**
Population growth outstrips resources

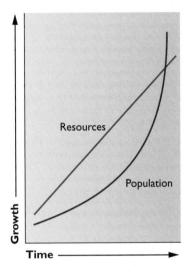

**(b) Boserup model**
Resources, inventions and technology keep pace with population demands

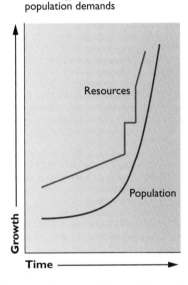

**Figure 30** Two models of the relationship between food supplies and demand

For decades, agricultural science has focused on boosting production through the development of new technologies. It has achieved enormous yield gains as well as lower costs for large-scale farming. But this success has come at a high environmental cost. Furthermore, it has not solved the social and economic problems of the poor in developing countries, which have generally benefited the least from this boost in production.

Any farming technique, whether on land or using fresh and marine waters, will have a negative impact on natural systems. This ranges from modification to total destruction. Farming may be seen as a system, with the inputs of technology creating detrimental outputs as well as food (Table 14).

**Table 14** Farming inputs and outputs

| Input | Purpose | Outputs — environmental impacts |
|---|---|---|
| Machinery, e.g. tractors and combine harvesters, and air transport for perishables | To replace human or animal labour, and increase efficiency | Pollution: increased fossil fuel use for transport and refrigeration. Increased food miles and packaging. Soil compaction and erosion. Loss of biodiversity in wetlands, hedgerows, forests. Trawling impacts on marine ecosystems and non-target species, e.g. dolphin, albatross |
| Chemical fertilisers, especially nitrogen products | To increase yield, by providing high levels of plant nutrients | Pollution: eutrophication of water by agricultural runoff, loss of biodiversity, e.g. sugar cane in Caribbean |
| Pesticides | To remove insects and other pests, which could reduce yields | Toxic chemicals, such as DDT, entering the food chain and causing damage to organisms, which were not the intended 'victim' of the chemical |
| Herbicides and fungicides | To remove weeds, which take up space and use up nutrients, and to reduce fungal diseases that reduce yields | |
| Antibiotics | To increase resistance to disease in livestock and fish and to increase yield | Antibiotic-resistant bacteria and the danger of epidemic outbreaks; fears for human health due to consuming meat |
| Animal/fish feed | To increase the density of animals/fish kept in a given area | Increased demand for food crops to be used to feed livestock, and therefore increased pressure to clear areas (e.g. forests) to produce more crops especially soya. Waste products from animal slurry are highly toxic. Methane from livestock is a potent global warming gas. Intensively reared cattle are fed diets rich in protein and energy. For every acre of feedlot in the UK two more are farmed overseas to meet its needs |

**Knowledge check 24**

What is the purpose of the many chemicals used as part of modern, intensive farming?

**Examiner tip**

Make sure you understand the differences between farming systems such as the Green Revolution, GM crops and organic farming in terms of their differing costs and benefits. The Green Revolution and genetically modified crops are often confused.

Any type of organic farming has relatively low impact on the environment compared to large-scale intensive or extensive commercial ventures. These have dominated modern farming in Westernised countries since the Second World War, and more recently developing countries have intensified agriculture with the Green Revolution. In the EU, reforms of the Common Agricultural Policy now foster more environmentally friendly agriculture, with a growth in LEAF farms (Linking Environment with Farming). However, organic farming tends to be less profitable and less productive than intensive chemical farming, and has suffered because of the global recession of the early twenty-first century. It must be remembered that the majority of farmers in developing counties are still rooted in subsistence and small-scale production methods, which are often also organic because of poverty.

# Enquiry question 3

# What is the role of desertification in threatening life at the margins?

## Scale and impact of desertification

**Desertification** is defined by the UN Convention to Combat Desertification (UNCCD) as 'land degradation in arid, semi-arid and dry sub-humid areas resulting from various factors, including climatic variations and human activities.' Land degradation means the reduction or loss of the biological or economic productivity of drylands. 70% of drylands are classed as degraded (excluding hyper-arid deserts) and suffering desertification.

**Drylands** include all terrestrial regions where the production of crops, forage, wood and other ecosystem services are limited by water. These are places where the climate is classified as dry sub-humid, semi-arid, arid or extreme hyper-arid (where few people live). The precipitation amounts are often unreliable and sporadic, which makes ecosystems fragile and vulnerable to over-use. Drylands cover about 41% of the Earth's land surface and are inhabited by over 2 billion people (about one third of world population). Desertification is found on all continents except Antarctica, and affects the food security of over 2 billion people. Africa and Central Asia are particularly at risk, with three key areas of vulnerability in sub-Saharan Africa: the Sahel, the Horn of Africa and southeast Africa. Severe droughts occur on average in Africa once every 30 years, although the major cause here is human mismanagement.

## Causes of desertification

The Millennium Ecosystem Assessment identifies the cause as the long-term failure to balance human demand for ecosystem services and the amount the ecosystem can supply. Pressure is increasing on dryland areas for food supply for people and animals, fuelwood, building materials and water. This increase is linked to a combination of factors shown in Figure 31.

## The characteristics and vulnerability of dryland ecosystems

Drylands are characterised by a water deficit. This constrains two major interlinked ecosystem services: primary production and nutrient cycling, which directly govern both natural and agricultural systems. Crop production, forage for animals and fuelwood supplies are affected. The level of aridity partly determines whether nomadic or sedentary agriculture is practised. In more developed regions sophisticated irrigation systems may allow more intensive agriculture, but most dryland regions have extensive forms of food production.

Desertification has both direct impacts on non-dryland ecosystems and indirect impacts on people outside the actual regions, through dust storms and advancing deserts. Droughts and loss of land productivity are major factors in creating environmental refugees.

> **Knowledge check 25**
>
> Explain the characteristics of the world's drylands.

> **Examiner tip**
>
> It is useful to produce your own, labelled world map of dryland locations and areas that are suffering from desertification. Do this on an A3 outline world map.

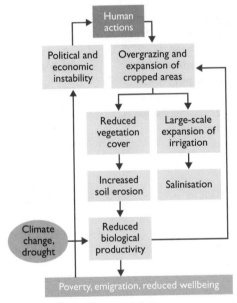

**Figure 31** Simplified causes of desertification

> **Knowledge check 26**
>
> Explain why the causes of desertification are complex.

> **Research activity**
>
> From sources like UNEP, the UNCCD and Greenfacts, build a factfile on at least two desertified areas from contrasting economic areas on the characteristics of dryland environments. Use headings suggested in Table 11. Ensure you include the vocabulary of: hyper-arid, mineral crusts, flash floods, net primary productivity, food chains, resilience, xerophytes and pyrophytes, role of producers, consumers and decomposers, e.g. termites. Key references: **www.unccd.int/** and **www. greenfacts.org/en/desertification/**

## Food production, supply and desertification

Food production generally requires massive amounts of water, for example to grow 1kg of wheat needs 1,000 litres of water, and to grow 1 kg of rice 3,000 litres. Irrigation can ensure an adequate and reliable supply of water, which increases yields of most crops by up to 400%. Although only 17% of global cropland is irrigated, it produces 40% of the world's food. Increasing irrigation efficiency and limiting environment damage caused by salinisation, damaged aquifers or reduced soil fertility are important for ongoing food availability. The sensitivity to human pressure increases with aridity. Human population in drylands increased by 18.5% between 1990 and 2000. Forty-three percent of land in Africa is classed as drylands, supporting 45% of the total population (325 million). Many African countries suffer from desertification and frequent recurrence of drought but are also vulnerable because of their poverty. Natural resources are critical for survival, land degradation occurs due to population growth, low technological development and unsustainable resource use. Many African countries suffer from weak institutional and legal frameworks and corruption, and lack the capacity to deal with the problems of desertification and drought.

# Enquiry question 4

# How effective can management strategies be in sustaining life at the margins?

## Strategies and techniques to increase global food supply and security

Increasing the food supply differs from increasing the security of supply. Both are difficult to achieve, but globally, increasing food supply has been given the priority. A **strategy** means the overall aims and tactics of a scheme, and it is implemented by various delivery **techniques** involving a range of policies and actual technology. Tackling insecurity means looking at all aspects of the food system from field to table and back again.

Figure 32 shows the stages required to improve food security and Figure 33 summarises the range of strategies that might be used in the quest for food security.

**Figure 32** Stages in tackling food insecurity

**Figure 33** Spectrum of strategies to tackle food insecurity

## Food security and trade liberalisation

Increasing food production leads to greater availability of food and generates income, which can help break poverty cycles. Trade liberalisation is opening up markets slowly, but there are costly barriers to overcome. The Doha round of talks by the World Trade Organization has so far failed to agree further trade liberalisation that might assist developing countries to participate in the global trade environment. Until there is fairer trade on a global scale, rather than the small but growing Fair Trade movement spearheaded by NGOs, food security is unlikely.

## Players and the need for more international efforts

There are a whole range of players from individuals to global institutions involved in the delivery of strategies (Table 15).

> **Knowledge check 28**
>
> Explain what is meant by Fair Trade.

**Table 15** Players in the delivery of strategies

| Player | Role in sustaining life of the margins | Examples |
|---|---|---|
| Individuals, e.g. farmers | Direct producers of food<br>Communities harbour stores of valuable local knowledge, coping strategies and innovation<br>Their cooperation is critical to ensure environmental sustainability | Role of Fair Trade organisations and impact on individuals<br>Local organic farms, LEAF |
| Government | Provides funding for agricultural research and development<br>Important in creating political and economic conditions to ensure stability of food supply<br>Response during times of crisis | Large-scale rehabilitation projects, e.g. China's Great Green Wall, Japan or UK overseas aid projects<br>Legal and conservation frameworks |
| Businesses and TNCs | Research and investment into new farming methods and technologies<br>Resource exploitation and trade in cash crops, fertilisers and farm machinery<br>Profit motive | The development of GM crop varieties such as Golden Rice<br>Monsanto TNC |
| NGOs and foundations | Community level support for farmers in the developing world<br>Education, training and skills providers<br>Many promote social equity, for instance female empowerment | Implementation of sustainable dryland farming in Sudan by Practical Action<br>Emergency aid, e.g. Oxfam<br>Toyota Motor Corporation<br>The International Alliance Against Hunger |

| Player | Role in sustaining life of the margins | Examples |
|---|---|---|
| Research organisations | Conducting scientific research on new crop varieties and farm systems<br>Not for profit motives<br>Education and skills training of farmers | The development of HYVs by the International Rice Research Institute<br>AGRA's work on a Green Revolution in Africa |
| IGOs such as UNEP and FAO | Promote international cooperation<br>Implement global actions such as the Millennium Development Goals<br>Monitoring and research to identify problems and seek solutions<br>Development assistance and aid to the developing world | World Bank's Global Response Food Programme<br>1994 UN Convention on Desertification |
| Watchdog pressure groups | Research and information gathering<br>Lobbying of agencies | World Resources Institute<br>USA Coalition Food<br>SUSTAIN |

## Sustainable strategies

There are three overarching aims if management is to be sustainable:
- supply of food should meet the likely increase in food demand
- food security for all
- environmental sustainability for long-term security

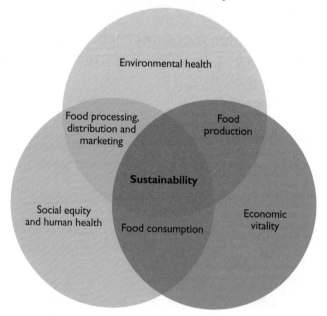

**Figure 34** The basic criteria for sustainability

Figure 34 shows the overlapping nature of the food system linked to the three basic criteria needed for sustainability. A well-functioning food system not only improves human health and social wellbeing, but also, in the long term is positive for the environment and the economy and allows them to cope with shocks from natural and human created disasters.

Overcoming food supply and security requires a range of different techniques. Table 16 shows combinations of techniques in any one policy.

**Table 16** Different techniques for overcoming food supply and security

| Immediate relief | Longer-term international response | Diversification |
|---|---|---|
| Examples: World Bank Global Food Response Programme; Reliefweb food parcels | Example: UN Convention to Combat Desertification helping prepare national strategies | Example: aquaculture: freshwater fish ponds for, e.g. Tilapia in Malawi; also important for protein |
| **Bottom up** | **Top down** | **Low technology** |
| Favoured by NGOs, often low technology more appropriate to local area, but may not cover a wide area | Where national strategies are run by government organisations — often involving prestigious high profile, high technology projects, e.g. megadams | Example: Sahel eco-farm approach: intercropping, microdosing of fertilisers into crop planting holes |
| **Technocentric** | | |
| From dams to e-technology. Increasingly innovative ways of using mobile phones in even remote and poorer areas: mobiles are used to find the best markets for perishable crops like bananas and fish. At an institutional level: UNICEF has had recent success with the text messaging Rapid SMS system in monitoring and delivering the protein-rich ready-to-use therapeutic food, Plumpy'nut, in drought-hit Ethiopia | | |

**Research activity**

Collect case studies from a range of organisations and scales, including more top down, often technocentric schemes like China's Green Wall afforestation, ICRISAT in India. Also bottom up, smaller scale schemes such as Sahelian eco-farms, Farm Africa, the World Fish Center. Evaluate their success in increasing food security by setting clear criteria.

Key reference: **www.droughtnet.org**

## Sustainability: eat well and save the planet?

Rather than the 'business as usual' scenario, with little thought for future generations, sustainability is about seeking ways of providing food, water and energy that are long-lasting and have less of an impact on the environment. Sustainability includes:

- all aspects of the food chain from production, transportation and processing to its delivery on to a plate
- 'local' — minimising energy used in food production, transport and storage

Places like the UK are effectively at the top of the global food pyramid, and play a vital role in developing global food security, especially in a century dominated by fears of climate change.

**Research activity**

How would the ethical and environmental policies suggested by pressure groups such as the Fairtrade Foundation and Sustain help achieve greater food security? Is there any evidence that governments are taking on such policies? You could look at the UK's Sustainable Commission and Department for International Development websites and the impacts of palm oil development for food and biofuels. You also need to investigate the increasing threat of climate change to sustainable food supplies and security. Starting points would be Oxfam, the Famine Early Warning System and World Food Summit websites. Key references: Food sustainability: **http://news.bbc.co.uk/2/hi/in_depth/629/629/7148876.stm** and **www.sustainablefood.com/**; NGO, e.g. Practical Action in dryland areas: **http://tinyurl.com/3kuyboq**

**Examiner tip**

Make sure you are aware of the difference between short-term, severe food shortages, which directly endanger lives (called a famine), and longer-term, chronic lack of food security, which leads to malnutrition and lowers life chances.

**Knowledge check 29**

Despite the UK being at the top of the global food pyramid, a small number of UK citizens are food insecure. Who are these people likely to be?

- You should research why some people suffer from undernutrition and malnourishment and why others suffer from overnutrition and obesity. Where are these issues most common and why?

- Which areas of the world are most food insecure, and how have trends in food security changed over time.

- What does being 'on the margin' mean for people in megacities, rural areas in Africa and other places where food insecurity is high?

- Explore how the drive to incease food production through ever more intensive farming has often led to the environment being damaged.

- Some areas of the world suffer from famine from time to time, whereas in others there are food surpluses and waste. Examine why this is and the complex economic, social, environmental and political forces that have led to these imbalances.

- Investigate the role of rising populations in adding to pressure on food supply and creating insecurity; the competing theories of Malthus and Boserup are important.

- You should research examples and case studies of a range of ways of increasing food supply such as the Green Revolution, GM crops, hydroponics, organic

- farming, aquaculture and others. It is important to be able to evaluate the costs and benefits of these.

- Investigate the global location of drylands and how they are used to produce food; this should include a consideration of the vulnerability and fragility of dryland ecosystems — are they suitable locations for food production?

- Link this to a study of spreading deserts and the complex human and physical causes of desertification; how much of a threat is desertification to food production?

- Research how politcial stability, careful land management and new technologies might help increase food production in food-insecure areas. What role does Fair Trade have in increasing food security?

- Contrast different types of technology that could be used to increase food supplies, such as intermediate technology versus more intensive farming methods.

- Consider the role of players such as IGOs and NGOs in securing food supplies, both in the short and longer term.

- Investigate whether or not sustainable food production strategies have a role to play in food supply increases.

# The world of cultural diversity

## Introduction

One world, close to 7 billion people by 2011, more than 200 nations, 6,912 living languages, many cultures and even more sub-cultures. One of the growth subdivisions within geography is that of **cultural geography**, the study of the characteristics of culture.

Here are some of the topics you will study in this option:
- Punk, Goth and Hip-Hop sub-cultures
- McDonaldisation and Starbuckisation
- Role of companies like Walt Disney and Microsoft
- Mall culture
- Compulsory Welsh

Since the mid-twentieth century, culture has changed faster than ever before with the growth and flows of technology, media and globalisation. All urban and most rural landscapes are the product of culture. The exploitation, vulnerability and management of landscapes, i.e. places, are often the result of cultural decisions. However, culture is a complex concept, which has different meanings to people in different places and changes over time. At present some cultures are embracing or localising globalisation, others actively resisting its associated uniformity.

**Figure 35** Topic concept map

**Figure 36** The synoptic context

**Table 17** Checklist for the world of cultural diversity

| Enquiry question 1 | Enquiry question 2 | Enquiry question 3 | Enquiry question 4 |
| --- | --- | --- | --- |
| **What is the nature and value of culture in terms of peoples and places?** | **How and why does culture vary spatially?** | **How is globalisation impacting on culture?** | **How do cultural values impact on our relationship with the environment?** |
| • Type of culture? Sub-culture?<br>• Type of landscape produced: ethnoscape? financescape? commodityscape? mediascape?<br>• Pressures on the culture and cultural landscape<br>• Is the culture/landscape vulnerable? How? Why?<br>• How is local culture valued? Protected? | • Factors influencing degree of cultural diversity/homogeneity<br>• Urban? Rural?<br>• Players involved and attitudes to cultural diversity or homogeneity<br>• Attitude of local and national governments<br>• Evidence of cultural imperialism? | • Significance of globalisation on area<br>• Role of global media corporations<br>• Evidence of globalisation? Hybrids?<br>• Link to global consumerism | • Type of attitude to environment<br>• Exploitation?<br>• Preservation?<br>• Management and mitigation?<br>• Role of green movement<br>• Efforts at sustainability |
| Key refs used | Key refs used | Key refs used | Key refs used |

# Enquiry question I

# What is the nature and value of culture in terms of peoples and places?

## Definitions

Boris Johnson, Mayor of London, said on the BBC programme *The Culture Show*: 'Culture is what distinguishes men, human beings, from animals'. On this show, celebrities debate current cultural issues, including: *EastEnders*, sex, coffee, football and binge drinking. This type of culture may be best described as 'popular' culture rather than 'high' culture, which might be represented by art and classical music.

The very word culture suggests growth and change. It is basically a **system of shared values** in a society, which influences lifestyles and creates boundaries for behaviour and interaction with others (Figure 37). These boundaries can be seen in dress and greeting codes, religious festivals, and also in architecture and use of cityscapes and rural landscapes. Some governments, like the UK, have a separate department devoted to the development of culture.

Cultures develop norms, values, and behaviours that are suited to that environment. Over time, they take on the strength of **tradition**. Even when circumstances change, traditions often do not, since early conditioning is very difficult to overcome.

Culture is dynamic:
- It is passed on from generation to generation.
- It evolves naturally over time, particularly with increased contacts with other groups and beliefs — although individuals may perceive that their culture has not changed.
- It is open to external influences, for instance from another culture. Newcomers to an area, whether from inside a country or migrants from overseas, will either resist the culture they join or assimilate. Even third generation immigrants may retain some of their original culture. This is a two-way process where the core culture will also be changed, as shown in food tastes, religion and sometimes clothing in areas rich with migrants.

**Figure 37** Components of culture

Cultures are affected by contact between societies. Contacts may be **voluntary**, as in economic migration and holidays, or **forced**, by war and refugee migration. Cultural ideas may transfer from one society to another, through:
- **Assimilation** — the adoption of a different culture by an individual or group, often quickened by intermarriage.
- **Diffusion** — the movement of cultural traits and ideas from one society or ethnic group to another. While the main form of a trait may be transferred, the original meaning may not, as in a McDonald's hamburger. These are considered cheap fast

food in the USA or UK but they are more of a special occasion food in China or India, where the veggie burger has globalised the original form. Stimulus diffusion means an element of one culture leading to an invention or propagation in another.

- **Acculturation** and **transculturation** — processes by which a culture is transformed due to the adoption of cultural traits from another society. Colonisation has replaced the cultures of many indigenous people, as seen in Australia where Western culture dominated the original Aboriginal way of living. The Aborigines fiercely defended their traditional culture and are now better respected.

Groups may reject the values and beliefs of a core culture. These groups fall into two categories:

- **Separatists**, who often choose to live in their own enclaves and minimise contact with others, for example employees of a TNC in a foreign country or 'sunseeker' retirees from Britain in Spain. More settled groups include the Amish in the USA and Mennonites in Belize. Cults are specific religious groups, often living together in a community with a charismatic leader. They are often seen as potentially dangerous, unorthodox and extremist by the dominant religious organisations in a society.
- **Sub-cultures** seek to challenge or undermine the core culture and are initially seen as shocking and dangerous, but watered down versions of their 'culture' often enter the mainstream eventually. Examples are punks in the 1970s, and the eco-warriors who campaigned against road developments in the UK in the 1980s.
- The expression of any culture is in its buildings, landscaping and attitudes to conservation and exploitation.
- Governments may foster and help maintain culture by investing in specific signposting around housing estates and tourist trails.
- Schools, community centres and even subways often have expressions of local culture in the form of painted murals, and more unofficial graffiti.
- Shopping malls have a distinct cultural expression, even if it is a cloned type of expression where a mall may look similar in India, Mexico or France.
- Different organisations, from TNCs to hospitals, schools to local authorities foster specific cultures made up of the assumptions, values, norms and tangible signs (artefacts) of organisation members and their behaviours — e.g. Japanese factories in the UK.

**Knowledge check 31**

Explain the ways in which cultures can change and evolve over time.

**Examiner tip**

Signs, logos and brands are usually designed to 'express' key cultural ideas, often very selectively. Consider this the next time you visit a mall or high street.

**Research activity**

Identify the aspects of culture that define the UK, and why Wales, Northern Ireland and Scotland have devolved power and separate identities from England. Look at the recent Life in the UK test for citizenship: **www.lifeintheuktest.gov.uk/**. Investigate British culture using Department for Culture Media and Sport website **www.culture.gov.uk**, including the 'Taking Part' survey at **www.culture.gov.uk/what_we_do/research_and_statistics/4828.aspx**

EU portal on culture in all member countries: **http://tinyurl.com/o6tza**

Assess how Islam is not a culture but a body of principles and universal values, which can be acculturated.

Choose a minority culture suffering persecution, such as the Roma, and identify how such a community can retain its culture despite being scattered globally: **www.rroma.org/**

# Cultural landscapes

Place, as opposed to just space, is an important component of everyone's life, and is heavily influenced by cultural values. Cultural landscapes can be:

- Historic — the remains of an ancient culture's landscapes.
- Modern — a 'new' landscape reflecting the culture of today.
- Mixed — a fusion of the ancient and modern, which characterises most places today, and acts as a focal point for both the inhabitants of an area and tourists, e.g. city walls, churches and beauty spots.

Historical and beautiful landscapes are valued by most societies, which may attempt to preserve them by awarding Conservation Area status. In cities, new buildings adjacent to old ones may be designed to replicate their main forms. This can be seen in new malls that may have external architecture replicating surrounding historic features. Some countries with shorter preserved histories covet the historic sites of others.

A key player, the United Nations, designates landscapes such as the Taj Mahal, Ayers Rock or Stonehenge as World Heritage sites. More natural landscapes such as Dorset's Jurassic Coast or the USA's Wild West have the same high-level designation. Even landscapes heavily influenced by people, such as the industrial and farmed landscape of the Lake District or the reforested and grazed New Forest, are fiercely defended. The latest National Park in the UK, the South Downs, has been created to protect rolling chalk hills, under threat from agricultural techniques and urban sprawl.

**Knowledge check 32**

Explain what is meant by 'globalisation'.

The process of globalisation has created more globalised forms of landscapes as well as opposition to the erosion of local identity. Cultural landscapes that cross national and cultural boundaries can be classified as shown in Table 18.

**Table 18** Cultural landscapes

| Ethnoscapes | Created ethnic landscapes: Curry Mile in Birmingham, Chinatown in London and San Francisco or the nine New Towns outside Shanghai modelled on different cultures, e.g. Canadian-inspired Feng Jing complete with maple trees, and Thames Town, which looks like a British village. Allotments are a current form of conservation of a local culture |
|---|---|
| Financescapes | Statuesque landscapes of corporate tower blocks and offices such as Canary Wharf in London, the 101 Building in Taipei and the Pudong financial district in Shanghai |
| Technoscapes | Centres of hi-tech industry, often developed with hi-tech architecture such as Putrajaya in Malaysia, Silicon Glen in California, Cambridge Science Park, UK or Grenoble in France. Technoscapes of the past may be preserved, e.g. Battersea Power Station in London and La Villette in Paris |
| Commodityscapes | Sites of consumption, such as the old buildings making up Knightsbridge in London, or the famous market square in Marrakech — Djemaa El Fna — now a World Heritage site. Westfield shopping mall, London, is an example of the global cloning of retail centres |
| Mediascapes | With clever marketing, American music, movies, television, computer games and software have 'gone global', influencing the tastes, looks, and aspirations of virtually every nation |
| Ecoscapes | Landscapes based on natural features, such as the Grand Canyon, USA, Antarctica or Wolong in China |

## Pressures on cultures

Cultural landscapes are constantly under environmental, social and/or economic pressures to either change or be preserved, as attitudes alter and values shift in society.

Cultural **genocide** is the deliberate destruction of a group of people or nation for political, military, religious, ideological, ethnical or racial reasons, such as occurred in Rwanda in 1994. **Ethnocide** means the destruction of the culture of a people, as opposed to the people themselves, as China has done in Tibet.

On a smaller scale, but more common, are urban regeneration projects involving forced moves, seen in cities as diverse as London, Shanghai or Mumbai. The development of the Olympic Village site in Stratford and Newham was controversial and hotly debated by some local people because it may destroy the existing community and culture.

For many isolated and 'switched off' areas and countries, tourism offers a tempting route to economic development. However, it may also threaten their cultural values and landscapes, as seen in Bhutan, Macchu Picchu in Peru, and the Masai people in east Africa. An extreme form of cultural change is prostitution for the tourist trade, which has made places like Bangkok infamous. Even in the UK, tourism has created cultural changes, as seen in the rise of second home ownership in the Lake District or Wales.

Technology creates another interesting pressure: automobiles, televisions, stereos, cellular phones, computers and iPods. These have made it easier to know about the values of people all over the world.

Sectoral change in employment has led to the demise of cultures associated with rural primary production. As farming has declined in the UK, leisure and tourism have increased in rural areas. Urban influences have escalated, from counter-urbanisation to the banning of fox hunting.

## Protection and players

One influential global player is the United Nations Educational, Scientific and Cultural Organization (UNESCO), founded in 1945 with the rather ambitious goal to 'build peace in the minds of men'. It basically promotes international cooperation on education, science, culture and communication among the 193 countries which are member states. One of UNESCO's aims is the 'preservation and promotion of the common heritage of humanity' by World Heritage sites. This raises difficult questions about:

- which cultures and cultural landscapes should be preserved
- who should decide how and where cultures are to be preserved and promoted: a disproportionate 50% of all World Heritage sites are in Europe and North America, reflecting the influence of Western culture in UNESCO decisions

### Research activity

Assess how some landscapes are seen as a commodity, some as a national symbol, and some as a threat.

Choose a country/area in which a distinctive language is being deliberately retained as part of a cultural initiative, such as Spain's Basque region, or Wales.

Assess the role UNESCO and the EU have on cultural diversity, e.g. a 2008 concert celebrating cultural diversity: **www.youtube.com/watch?v=9flFbJw3pks**

EU Commission for Culture: **http://ec.europa.eu/culture/index_en.htm**

UNESCO: **www.unesco.org/culture**

### Research activity

Research how different groups of people can vary in their views about cultural landscapes (Table 18) and value them very differently. See the work of a New York-based anthropologist: **http://tinyurl.com/25gfk5**

### Examiner tip

Forced cultural change, through genocide and other means, was common during the colonial era. It is rare today.

### Knowledge check 33

What is counter-urbanisation and how could it be a force for cultural change?

### Research activity

Assess how culture may affect health. New diseases from new contacts, HIV/AIDS, anorexia, female mutilation, binge drinking, smoking, are all linked to cultural traits. A starting point is: **www.answers.com/topic/cultural-factors** and **www.york.ac.uk/healthsciences/equality/cultural.htm**

### Knowledge check 34

What is a UNESCO World Heritage Site?

# Enquiry question 2

# How and why does culture vary spatially?

## Homogeneity and cultural diversity

The amount of cultural diversity varies greatly across the world, at both national and local levels. Some countries are more culturally homogenous than others, such as Japan, Iceland and Saudi Arabia.

In contrast are the large-scale 'melting pots' of countries such as the USA, with large immigrant populations, and consequently a wide range of religious beliefs, customs and values. With successive generations the original culture may be diluted. Differences still remain between rural and urban cultures: African American from European, East Asian from South Asian, and religious believers from secularist.

**Research activity**

Contrast and compare the following policies on cultural diversity. UK government site: **www.homeoffice.gov.uk/**. Blog and information site for foreigners in the UK: **www.foreignersinuk.co.uk/contacts_119.html**. Japan's culture and migration policies: **www.migrationinformation.org/Profiles/display.cfm?ID=487**. And conversely, the cultural assimilation of Japanese immigrants into Brazil: **www.nowpublic.com/culture/100-years-japanese-immigration-brazil**. The TNC Nissan gives an interesting viewpoint on cultural diversity: **http://tinyurl.com/446jph8**

## Cultural diversity in rural and urban areas

The key 'switched on' locations in the world are certain cities where success has bred success. These gateways or hubs automatically attract not only large numbers of immigrants but also the youthful and vibrant people most eager to assimilate change. Hence cultural diversity is inevitable, as seen in two key world cities: London and New York, or the megacity of Los Angeles.

Rural areas globally tend to have lower ethnic diversity because of fewer employment opportunities and greater isolation, especially those areas that are 'switched off' from our globalised world. Rural communities may be less welcoming to different cultures, and whether rich or poor, immigrants are often classed as outsiders. Grants are available in many countries, such as the UK, to improve cultural facilities especially based on the arts, e.g. the Rural Development Council.

**Figure 38** Factors affecting cultural diversity

**Examiner tip**

In your report, it is important to use the language of culture such as *diversity*, *homogeneity*, *melting pot*, *assimilation* and *globalisation*. This will help make your report precise and focused. Compiling a glossary is a useful way to help learn the meaning of terminology.

**Knowledge check 35**

Explain the difference between the terms 'world city' and 'megacity'.

Investigate the causes and consequences, including the spatial distribution, of cultural diversity in a world city such as London compared with smaller cities and rural areas. Key references: multiculturalism — Joseph Rowntree Research Foundation: **www.jrf.org.uk/ knowledge/findings/housing/1950.asp**. Interculturalism using London and Bristol as specific case studies: **www.guardian.co.uk/graphic/0,5812,1395103,00.html**; **http:// tinyurl.com/3hc695v**

## Attitudes towards cultural diversity

Not all cultural and political groups agree that cultural diversity is positive. Tolerance of others is not universal and the host country may reach a tipping point when the carrying capacity, either real or imagined, is reached. Recent changes to immigration policy in the UK suggest this threshold has been reached. Attempts to help assimilation of other cultures into mainstream culture may be seen in education policies.

Diversity may also be seen as a threat to local culture, as happened in Spain when Franco was in power and banned the use of the Catalan, Basque and Galago languages. Regional languages are seeing a revival with promotion of diversity, e.g. Breton, Welsh, Gaelic and Catalan.

## Cultural imperialism

This is the promotion or imposition of a more dominant culture on to another culture. This has happened throughout history after war and conquests of new territories — most recently in the twentieth century after the dismantling of the old colonial system and a rise in Western, especially European and North American, power.

McDonaldisation is the term given to the extreme form of consumerist culture originating from the USA. Western culture generally has:
- an emphasis on high levels of resource consumption
- a high reliance on globalisation
- a political system based on 'democracy', although in differing forms
- a 'technological fix': a reliance on, and belief in, technology
- a high degree of influence from global media corporations (BBC, CNN), television from the USA, Hollywood films and global brands like Coca-Cola and McDonald's.

Supporters of the Western cultural model would describe this as modernisation or development, and a belief that this culture represents progress, rather than cultural imperialism.

Counter movements from other cultures, especially green movements, and some religious groups would disagree, as shown in demonstrations at every World Trade Organization or G8 and G20 meetings.

**Knowledge check 36**

Explain why mass tourism and mass immigration might be seen as a threat to the survival of cultures.

**Research activity**

Contrast the policies of the UK and France on immigration assimilation. Tibet and Israel provide contrasting case studies of viewpoints on cultural landscapes: **www.international. ucla.edu/article. asp?parentid=2732 http://mondediplo. com/2008/05/09tibet**

**Examiner tip**

Because most of us live within Western culture, it can be hard to identify its key features. Try discussing what is important in your life with someone from a different religion, ethnicity or area (e.g. rural, if you live in a city) to try to identify differences in values and perspectives.

Research the mechanisms of the process that has become stereotyped as 'McDonaldisation' and the counter-cultural movements that it has given rise to: **http://www.tuc.org.uk/theme/index.cfm?theme=globalisation**; **www. thehumanist.org/humanist/articles/essay3mayjune04.pdf**; **www.wsu.edu/~amerstu/pop/cultimp.html**

# Enquiry question 3

# How is globalisation impacting on culture?

## Viewpoints on globalisation and cultural diversity

There is a debate between those who feel that globalisation will ultimately destroy cultural diversity and those who are sceptical about its power to do so (Figure 39).

Figure 39 The globalisation debate

## The role of global media corporations

Media corporations have become major players in spreading certain cultural values, and many would argue that they have disproportionate power. The companies involved, such as Disney, Viacom, Microsoft, Vivendi, argue that no one is forced to buy, watch, listen to or use their products. However, they do monopolise the view of the world for billions of people, and a matter of concern is that recent mergers mean less competition and a uniformity of coverage even to the point of reducing the diversity of languages.

## Globalisation and hybrid localised culture

The idea that globalisation will produce one global culture is too simplistic. New cultural forms and landscapes as well as new languages are emerging as economic migrants increase and as the internet and texting become common even in remote areas. There are now distinctive hybrid globalised forms of fashion, music and film.

Modern technology can be seen not only as a threat to traditional forms of culture but also as a cultural opportunity. Websites such as YouTube have encouraged the global village in a virtual medium. Free social messaging websites like MySpace and Facebook, together with blogs, and the availability online of television, news and film, mean that it is possible to keep in touch with family and friends and expand cultural contacts far more easily and over a wider spatial extent than ever before. This may work to preserve at least some aspects of cultural diversity.

## The impacts of global consumerist culture

Increasing wealth and globalisation may increase choices, from basic food types to leisure pursuits. Indeed, many poorer people may choose the outward signs of globalisation for prestige and a feeling of belonging to mainstream society.

However, consumerist culture may not be best for mental or physical health. Too much choice, whether in information or brands of trainer, mobile phone or products like milk, may lead to more stress than satisfaction. Research in *Scientific American* in 2004 indicated that in affluent societies like the USA, GDP has more than doubled in the past 30 years yet the percentage of people describing themselves as 'very happy' has declined by about 5%.

# Enquiry question 4

# How do cultural values impact on our relationship with the environment?

## Cultural attitudes and the environment

The United Nations defines the environment as, 'the totality of all the external conditions affecting the life, development and survival of an organism' which could mean a tree, polar bear or human.

Environment can be divided up into urban, semi-urban, rural and wilderness areas.

Different groups have different perceptions and place differing degrees of importance on the environment. With the present growing threat of climate change, people view even the atmospheric environment differently depending upon the country, organisation or individual. Many view environmental issues as a global, governmental problem rather than an individual, local one, hence the emphasis of the Rio Earth Summit of 1992 on its Local Agenda 21 of 'think global, act local'. Cultural views can vary from sensitive to indifferent.

Several factors affect current cultural attitudes to the environment:
- rising net income
- increased education
- personal involvement/impacts from pollution, soil erosion, etc.
- rise of local branches of international green groups such as Greenpeace, WWF
- role of media: television for information, and mobile phones and the internet to mobilise behaviour
- role of government in disseminating information and allocating resources to environmental protection: from pollution control to National Park designation. The international image of a country also now depends on its attitude and effectiveness in environmental quality, especially as foreign direct investment and TNCs can be very selective in such a globalised world
- role of IGOs, e.g. UNESCO and UNEP in World Heritage Site and Biosphere Reserves

By the start of the twenty-first century there were four main cultural views of the environment:

Which two of the four cultural views of the environment would require humans to radically change their behaviour if they became culturally widespread?

**Research activity**

Investigate the recent changes in attitude to the environment in China with the Beijing Olympics 2008 as a possible catalyst: **http://tinyurl.com/3j7tc49**

Summarise the main findings of the EU Eurobarometer survey in 2008 about attitudes and knowledge of the environment and global warming: **http://tinyurl.com/3ejnkm**

- **Technocentric view** — the historic Western idea that the environment is a source of resources for humans to use and a sink for waste.
- **Accommodation view** — the view of most Westernised countries at present. Improve the existing system through finance, technology, smart techniques — no fundamental change.
- **Social ecology view** — a new system is needed with community and local focus, ethically maintained, e.g. recycling, organic farming, solar energy. Ecotowns lead the way.
- **Deep ecology view** — going back to pre-industrial values. Nature dominates; careful human use of natural material; humans are part of the environment.

These viewpoints will probably all be represented in large democratic countries and may well change over time. People and societies constantly change in their cultural outlook.

Many governments, even originally exploitative ones like the UK and Australia, are moving closer to environmental cultural views that enhance ecological awareness, urban lifestyle quality and fairer societies.

China, the world's most populous state is a huge consumer of resources and pollution generator. At present, environmental attitudes in China are shaped first by tradition and culture, and second by new conceptions of economic growth and consumption. There is a cultural history of humans trying to control nature, and exploit it. But there are signs of a backlash against the immense environmental degradation resulting from China's rapid economic transformation into an industrial superpower.

## Valuing landscapes

A major part of the environment is landscape, which may be valued for a series of reasons. As mass tourism and eco-visits increase, protected landscapes are increasingly seen as a marketable attraction with spin off multiplier effects in the economy.

People's cultural perceptions of landscape value determine how land and resources are used or abused:
- landscape as sacred and preserved
- landscape as pleasure — conserved, but may become degraded and polluted
- landscape as life and protected
- landscapes have resources to use for profit and can be degraded and even destroyed

Protecting areas is no longer the preserve of more affluent societies that can afford to choose not to exploit an area and protect it as a nature reserve or National Park. The first National Park was in the USA — Yellowstone, 1872. Worldwide there are now National Parks in almost 100 countries and China has 187 — all designated since 1982. Most include some degree of use as a leisure hot spot and provide a 'natural' cultural outlet. See: **http://tinyurl.com/6ng63z**

Figure 40 shows a model based on the Kuznet environmental curve of how attitudes vary over time to the protection of the environment.

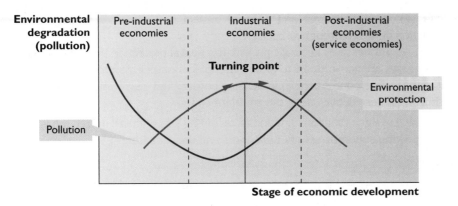

**Figure 40** Attitudes to environmental protection

# Sustainable environments

Sustainability is difficult to define, and is used by different cultures in different ways. The concept began with late twentieth century concerns over misuse of the Earth's resources and the increasing development gap. Sustainable development essentially seeks to generate economic progress that is equitable, will last, involves ordinary people in decision making and does not needlessly damage the environment.

According to the UK Sustainable Communities Plan: 'sustainable communities are places where people want to live and work — now and in the future. Culture can play a role in building sustainable communities.'

However, sustainability is often used as more of a 'green' concept focusing on environmental issues rather than social equity and public participation. Globally, local authorities and businesses stress 'environmental sustainability'.

The biggest global organisation setting standards to encourage universal agreements on ethical issues is UNESCO. It has set conventions to address the fact the 'world urgently requires global visions of sustainable development based upon observance of human rights, mutual respect and the alleviation of poverty'. The EU is a signatory, as is China.

# Consumer culture

Parents have an in-built desire to ensure a better lifestyle for their children. In Western culture this has developed into a belief that greater material wealth means more secure and happier people. Counter-cultures such as the eco and environmental movements dispute not only this but also the anthropocentric values underpinning consumerism — that humans should dominate the planet because they are more important than anything else.

# Environmentalism versus consumer capitalism

Most people now live in a world dominated by capitalism, and paradoxically the same people who are profligate consumers are concerned about environmental issues — global warming, increased hydro-meteorological hazards like desertification, and biodiversity loss.

**Knowledge check 40**

Explain why more affluent, post-industrial societies seem to protect, and perhaps value, the environment more than less developed industrialising ones.

**Knowledge check 41**

Explain the difference between sustainable development and environmental sustainability.

**Research activity**

Look at the UK government's official policy on sustainability and contrast it with that of a different cultural tradition, e.g. China or Morocco: **www.defra. gov.uk/environment/ economy/ sustainable/**

Why do big TNCs like Shell or companies like M&S market themselves as 'green'? Can they ever be really green?

Research UNESCO's viewpoint on the role of culture in sustainability.

**Examiner tip**

It's worth remembering that back in the 1960s, environmentalism was a fringe counter-culture, which has gradually become mainstream culture.

Capitalism is an economic and political system that almost inevitably leads to destruction of the environment because:

- Short-term timescales for profit put environmental protection in second place.
- Businesses are accountable to their shareholders, rather than to the wider public.
- It marginalises poorer people, increasing poverty and desperation, which leads to short-term destructive use of the environment.

There are few signs that consumption is becoming less popular, indeed it has spread to the growing superpowers of China, India and the Gulf States.

**Research activity**

Investigate the growing demands of countries like China and poorer developing countries on global resources. Examine the conflicts between environmentalism and consumption by finding out how contraction and convergence could help reduce ecological footprints.

**Summary**

- Consider different ways of defining culture and different aspects of culture.
- Research the characteristics of a variety of different cultures, in contrasting locations, as well as the landscapes produced by those cultures (both urban and rural, and traditional and modern).
- Examine the vulnerabilty of cultures that are under pressure socioeconomically, from environmental change and political forces.
- Research examples of cultures that are protected and valued, as well as examples of those that are under threat and valued less. Consider the role of players such as governments in protecting cultures, or failing to protect them.
- Research contrasting cultures, some of which are diverse and others that are more homogenous, and examine the reasons for the contrast.
- Examine cultural diversity in cities and rural areas — fieldwork could be used to assist this.
- Consider the extent to which cultural imperialism exists and how it might manifest itself.
- Know what the process of globalisation is and how it might impact on cultures in both positive and negative ways; this should include understanding the role of media organisations in driving cultural globalisation as well as the development of new hybrid forms of culture and global cultural movements.
- Consider the impact of a global consumer culture on landscapes and cultures (especially more traditional ones).
- Using examples of contrasting cultures, examine how different cultures view the environment — both locally and globally; also examine the extent to which cultures protect or exploit the environment.
- Assess the view that the primacy of humans over the environment is necessary to justify the current consumer culture, especially in the 'West'.
- Research examples of cultural movements that are pro-environment and/or anti-consumption and consider their role in resolving the tensions between consumer culture and the need for greater environmental sustainability and cultural protection.

# Pollution and human health at risk

## Introduction

The largest global organisation devoted to health risk is the UN World Health Organization (WHO), which stresses: 'Better health is central to human happiness and wellbeing. It also makes an important contribution to economic progress, as healthy populations live longer, are more productive, and save more'.

**Health** is 'a state of complete physical, mental, and social wellbeing and not merely the absence of disease or infirmity' (WHO). There are many complex interrelated factors influencing **health risk** (Figure 41), which include the threats that cause both disease and disability or **morbidity**, and **mortality** (death).

Critical is the level of socioeconomic development and personal status, which will influence environmental conditions and how the risks are managed. **Global health risk** means health problems that are so large they have a global political and economic impact or **burden**.

**Knowledge check 43**

What is the WHO and what is its role in health and pollution?

**Figure 41** The health risk equation

You will study the patterns and trends in health risks over time and space, both globally and locally, and evaluate the factors involved, including globalisation, global shift of environmental pollution, management and players. Some of the work you have done at AS on Going Global will help here.

**Figure 42** Topic concept map

**Research activity**
When you research case studies and smaller examples keep notes on all the enquiry questions, with a log of sources that you can review once the research focus pre-release material is available. You will be expected to quote some key sources in your actual exam: from books, publications, newspapers, websites, DVDs, podcasts, pamphlets etc. Create your own graphs and comparisons between countries. Key reference: WHO, with factsheets, databases: **https://apps.who.int/infobase/**

One key point to remember when researching this option is that you must apply a geographical focus. The specification suggests that a synoptic view of the people and power involved in any health risk must always be linked to place (Figure 43). Do not get sidetracked by medical details unless they directly link to geography.

**Figure 43** The synoptic context

**Table 19** Checklist for pollution and human health at risk

| Enquiry question 1<br>**What are the health risks?** | Enquiry question 2<br>**What are the causes of health risks?** | Enquiry question 3<br>**What is the link between health risk and pollution?** | Enquiry question 4<br>**How can the impacts of health risk be managed?** |
|---|---|---|---|
| • Short-term or longer-term risk?<br>• Infectious/transmissible or chronic/degenerative disease?<br>• Pandemic, epidemic or endemic health risk?<br>• Geographical pattern at global/national/local scales?<br>• Temporal pattern<br>• Where on the epidemiology model and WHO transition model?<br>• Social, economic, environmental, long- and short-term impacts | • Range of causes<br>• Long-established (older) risk or emergent or re-emerging risk?<br>• Relationship to socioeconomic status<br>• Is it classed a health shock?<br>• Links to any geographical feature/pathway<br>• Influence of environmental change<br>• Model, e.g. diffusion | • Type of pollution?<br>• Source?<br>• Direct and indirect effects?<br>• Incident or sustained pollution<br>• Link with economic development<br>• Where on models — environmental. Kuznet curve, epidemiological model/WHO, transition model?<br>• Evidence of pollution fatigue in reducing health risk | • Socioeconomic and environmental impacts: short- and long-term burdens<br>• Management strategy/ policy — why used and how effective?<br>• Agencies involved<br>• International efforts and aims<br>• Role of long-term sustainable or short-term emergency relief |
| Key refs used | Key refs used | Key refs used | Key refs used |

Note: enquiry question 3 will only apply if the health risk is linked to pollution, unlike risks such as diabetes and HIV/AIDS.

# Enquiry question I

# What are the health risks?

An incredibly diverse range of health risks reduces quality of life and longevity. The key to success in this option is to choose a range of health risks across different areas of the world to identify the geographical features and factors involved.

**Table 20** Simple health risk classification (excluding genetic reasons)

| Infections, often communicable, often acute health risks, i.e. rapid onset or intense symptoms, split into: | | Degenerative, chronic health risks often resulting from longevity, not communicable: |
|---|---|---|
| • endemic, meaning a disease that has persisted for a long time in an area<br>• epidemic, when a disease fluctuates over time<br>• pandemic, when epidemics affect wide geographical areas | | • chronic (lasting over 3 months)<br>• cardiovascular (heart) diseases<br>• respiratory diseases, chronic obstructive pulmonary disease (COPD)<br>• obesity-related<br>• diabetes, cancer, depression<br>• maternal/perinatal illnesses<br>• degenerative<br>• arthritis, Alzheimer's |
| **Vectored, i.e. involves a carrier:**<br>• malaria or dengue from mosquito<br>• Lyme disease from ticks<br>• trypanosomiasis from tsetse fly<br>• plague from rats | **Non-vectored, i.e. through bodily fluids, food, airborne, waterborne:**<br>• HIV<br>• tuberculosis (TB)<br>• norovirus<br>• typhoid<br>• avian influenza | |
| **Pollution created/related health risks:** cholera, radiation, asthma, respiratory infections, melanoma | | |
| **Traumas:** e.g. from work related accidents or transport accidents | | |

## Causes of mortality and spatial patterns of health risk

Globally there are huge health inequalities, nationally and locally, both in terms of health during life and age of death (longevity). One common measure of the level of people's health, created by WHO is DALYs (disability-adjusted life years) — the years of life spent with reduced functions resulting from health conditions of varying severity.

**Knowledge check 44**

Name two infectious diseases and two degenerative health risks.

**Table 21** Globally — what people die of based on national wealth

| High-income countries (US$11,456 person⁻¹ yr⁻¹) | Middle-income countries | Low-income countries (US$935 person⁻¹ yr⁻¹) |
|---|---|---|
| • Low mortality countries<br>• 66%+ of population live over 70 years<br>• Chronic diseases dominate: cardiovascular disease, chronic obstructive lung disease, cancers, diabetes or dementia. Lung infection is the only leading infectious cause of death<br>• e.g. UK, USA | • 50% live to 70 years<br>• Chronic diseases plus HIV/AIDS, complications of pregnancy and childbirth and road traffic accidents dominate<br>• e.g. China, Brazil | • High mortality countries<br>• 25% reach 70 years<br>• 33% of all deaths are of under 14 years. Cardiovascular diseases are the main single cause of death but infectious diseases combined (especially HIV/AIDS, lung infections, tuberculosis, diarrhoeal diseases and malaria) cause more deaths. Complications of pregnancy and childbirth also important in mortality<br>• e.g. Kenya, Bangladesh |

Globally, one in three deaths is from infectious or communicable diseases such as HIV, concentrated in poorer regions often linked with malnutrition. However, the biggest threat in developed countries is from chronic, non-communicable diseases, especially cardiovascular disease. The speed at which many countries have changed in their socioeconomic structure over the past few decades means that health managers may have to deal with both types of health risk in the same country. WHO divides the world into high and low mortality regions, correlating strongly with industrialisation and GNP/GNI (Table 21).

**Knowledge check 45**

State how health risks differ between high- and low-income countries.

Use the economic development spectrum studied in AS Going Global:

MEDCs–NICs–LEDCs–LDCs

to analyse spatial variations. Starting points are WHO's annual global burden reports. For interesting graphs and podcasts see the Gapminder website.

## Health and wealth

There is an overall positive correlation between global income and health, with some anomalies like Cuba (low GNI but good health).

Cuba is an interesting country to find out about, as it has a health profile close to that of the developed world, but an income profile of a developing country.

Fifteen per cent of the world's population live in high-income countries, mainly North America and Europe, but these only account for 7% of all deaths annually. However, the simple North–South divide of the 1980s has shifted with the rapid rise of transition economies like China and the countries of Eastern Europe. By 2001 most people lived in middle-income countries, with the biggest gap ever seen between rich and poor and associated life expectancy and health risk. Since the 1990s, sub-Saharan Africa has regressed, mainly because of HIV/AIDS. Health inequalities show an inter-generational cycle, with differences beginning before birth. The scale of the problem is immense: with acute risks for the 1 billion people who live in urban slums.

Is life expectancy the same in all geographical areas of a developed country such as the UK or USA?

Apart from national variations, there are also great regional and local variations: most cities show differences in wealth and economic welfare. In Bristol, average life expectancy may vary by 10 years between affluent and deprived neighbourhoods. Imagine the differences between an affluent gated community in Mumbai or Rio de Janeiro and a newer shanty town.

Health inequalities depend on:
- demography: age, gender, age of child bearing
- environment: housing and pollution, especially water quality
- lifestyle and habit choices: opportunities, diet, education, hygiene, job choice, smoking, drugs, drinking, exercise
- healthcare services and management: medical and infrastructure technology, access and effectiveness of regulation

## Health risk patterns over time

Temporal patterns of health and disease (how health risks alter over time) are described by models like the epidemiology model (Table 22).

Which stage of the model in Table 22 does the 2009 swine flu pandemic (which is estimated to have killed 18,000) best fit into?

The length and characteristics of the model's stages vary with local and national social, cultural, economic and health resource factors. The fundamental shift from communicable to non-communicable diseases is predicted to be worldwide by 2020 (WHO). It began in the 1960s in low and middle income countries, with the fastest changes in 'switched on' places like South Korea, which used the experiences and more advanced technology of older industrialised nations. Stages 1 and 2 have often become compressed.

**Table 22** Epidemiological model based on Omram 1971

| | Stage or age | | | |
| --- | --- | --- | --- | --- |
| | **1 The age of pestilence and famine** | **2 The age of receding pandemics** | **3 The age of chronic diseases** | **4 The age of emerging/re-emerging infectious diseases** |
| Causes of health risk | Large number of infectious, acute diseases | Reduction in the prevalence of infectious diseases and fall in mortality | Degenerative and human induced diseases of affluence suffered by ageing populations | Emergence of new infectious diseases, e.g. HIV/AIDS or the re-emergence of 'old' diseases, e.g. TB, measles |
| Examples of types of health risk | Mainly respiratory and infectious diseases: measles, smallpox, malaria, typhoid, cholera, tuberculosis, enteritis, diarrhoea, pneumonia | | Cancers<br>Respiratory diseases including asthma | HIV/AIDS<br>SARS<br>Avian influenza |
| Link to pollution | Localised pollution, especially waterborne | Rise in all types of pollution as industrialisation increases | Environmentally conscious but consumerist society, which has pressurised governments into reduction in water and land pollution, but rise in air pollution | In low to middle income countries high rates of all types of environmental pollution |
| Link to economic development | Low-income countries, e.g. UK in 17th century.<br>Currently Ethiopia, Bangladesh, although most moving to second stage | Industrialisation, e.g. UK in 19th century. Currently low to middle income countries, e.g. India, China, especially in rural western provinces | Post-industrialisation, e.g. UK in 20th century. Currently upper income countries, e.g. UK, USA, plus recently industrialising countries like urban eastern China. Ageing populations in urbanised societies | Low to middle income countries less able to cope with the double burden of health risk, i.e. infectious and chronic diseases. Early 21st century: huge rise in HIV/AIDs, smoking, hypertension, toxic effects of widespread environmental pollution not under control |

The fourth stage was added because of several unexpected events during the late twentieth century:

- The emergence of HIV/AIDS in the 1980s has seriously hindered the shift, producing a double burden of disease affecting rich (USA) as well as poor (African) societies.
- In more advanced economies, major advances in the prevention and management of heart disease and many cancers, and remarkable increases in life expectancy for the over-50s, may be called a 'post industrialisation' phase — particularly in Westernised countries.
- In the former USSR (now Russia) the demise of communism in the early 1990s led to a collapse in public health systems, an increase in tobacco and alcohol abuse and a re-emergence of infectious diseases like diphtheria and tuberculosis — supported by increasing inequity and poverty.

**Knowledge check 48**

Explain why male life expectancy in Russia fell from 65 years in 1990 to 63 years in 2010.

**Research activity**

Models are a useful way of summarising complex reality and helpful as a basis for comparison. You should establish where your case study is located on the model and how it may change in the future. Refresh your knowledge on how sanitation, hygiene, medical advances, etc. influence death rates as shown in the demographic transition model.

Projections are vital for health system managers and players involved in resource allocation and overall strategies. WHO forecasts that by 2030:

- Chronic disease will be dominant, even in low income countries like Bangladesh, with decreases in infectious diseases like diarrhoea, HIV/AIDS, tuberculosis and malaria.
- There will be an increase in mental illness, especially depression, which causes disability rather than death.
- Vehicular accidents may rise to third in importance, followed by obesity and smoking.

## Health risk and effects on quality of life and economic development

Health risk causes a reduction of 'human capital':

- direct effect on the individual: poor health, shorter life expectancy, negative effects on children in family
- indirect effect on local/national economy: loss of work time and productivity and hence lowered GNI

# Enquiry question 2

# What are the causes of health risks?

## Causes and socioeconomic status

The causes of more established health risks, such as cancer; emergent health risks such as HIV/AIDS; and re-emerging health risks, such as tuberculosis, are complex because they include:

- internal causes, including genetics, lifestyle choices by individuals, employment
- external causes, e.g. local environmental conditions such as polluted rivers, governmental strategies, role of TNCs

Figure 44 is a model commonly used by health authorities to show the direct factors (inner rings) and more indirect factors (outer rings) affecting health.

Socioeconomic status is how an individual's economic ranking relates to others in their society. It directly affects morbidity and mortality, although the relationship may depend on factors such as gender, culture and ethnicity as well as management. It varies at global, regional and local scales.

**Examiner tip**

Take care with your spelling of diseases such as 'diarrhoea' and 'influenza' — remember that quality of written communication forms part of the DRACQ mark scheme.

**Research activity**

This section is effectively a foundation for the rest. Once you are secure in knowing the background, the key terms and the key players, then the rest will fall into place. When justifying your focus in your final report, show the huge range of health risks and, as a geographer, why you are concentrating on certain types and aspects.

**Examiner tip**

Take care not to over-play the impacts of extreme events, such as those shown in Table 23. Even very major natural hazards usually have a very small impact on overall death rates and health trends.

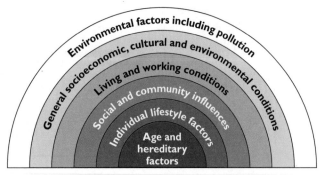

**Figure 44** Direct and indirect factors affecting health

The factors behind the initial development and subsequent spread of disease may be split into socioeconomic and environmental categories (Table 23).

**Table 23** Factors in the development and spread of disease

| Socioeconomic factors | Environmental factors |
|---|---|
| Wealth, which often controls access to:<br>• nutrition<br>• better sanitation and water standards<br>• medical and general infrastructure technology<br>• low risk employment<br>• higher levels of healthcare, especially immunisation<br>• better education, and improved status for women —'social vaccine'<br>Migration and exposure to health risks<br>Overuse of antibiotics and agricultural chemicals<br>Inadequate or deteriorating public health infrastructures | Environmental changes caused by human activity, e.g. dam and road building, deforestation, irrigation, and, at the global level, climate change<br>Natural hazard shocks, e.g. extreme heat waves and cold snaps, earthquakes and tsunami<br>Pollution |

**Research activity**

Health authorities within the UK have readily available data online, possibly mapped using GIS. Research the health status of your local area. See the Association of Public Health Observatories: **www.apho.org.uk/default.aspx**

## Geographical features and pathways

Geographical location is a major factor in the development, spread and consequences of health risk (Table 24). Infectious diseases are spread by pathogens (agents that cause disease) like viruses (e.g. measles and influenza), bacteria (e.g. meningitis and tuberculosis) and other microorganisms like parasites (tapeworms and guardia). These invade the body or contaminate the environment.

**Knowledge check 49**

Explain what is meant by the term 'pathogen'.

**Table 24** Geographical pathways and features

| Physical environment | Human environment |
|---|---|
| • Climate, topography, hydrology, geology, ecology. These may act as a barrier or conduit, or create a 'disease reservoir' for pathogens and vectors. They can be physical, e.g. lakes, or within humans themselves<br>• Humid tropical and equatorial regions produce optimum conditions for many infectious diseases. Unfortunately, often correlated with poorest, most vulnerable countries<br>• Changes in environment: global warming impacts | • Infrastructure: transport routes, especially air travel<br>• Services: sewage and fresh water<br>• Pollution pathways<br>• Building and environmental quality including green spaces<br>• Race, gender |

**Research activity**

Malaria and schistosomiasis, which involve vectors, are particularly good for showing the role of physical conditions and pathways. Meningitis, tuberculosis and measles show the role of human environments, and HIV/AIDS and SARS show communications/transport influences. Key reference: WHO website: **www.who.int/mediacentre/factsheets/en/**

**Knowledge check 50**

Which of the geographical pathways and features in Table 24 could globalisation affect the most?

## Emergent and re-emergent infectious diseases

There are repeated cycles in the rise and fall of infections. Currently there is a high rate of re-emergent diseases (tuberculosis and measles), and emergent diseases (Ebola fever, Legionnaire's and Lyme diseases, toxic shock syndrome, AIDS, hepatitis C, vCJD and a new variant of cholera).

You could research the 2009–10 HINI swine flu pandemic as it proved to be a major event, with widespread public health consequences.

Research activity

Models are used by managers to control health risk patterns and GIS is an important tool. Other models you could investigate are by Wilkinson (life expectancy/income) and Bartlett who used measles to show how a disease can be endemic (remain in a population as a 'reservoir' of infection — UK, 2008). Key reference: Science Museum — Making the Modern World: **http://tinyurl.com/y9cvlyo**

There has also been an increase in the virulence and drug resistance of infections such as diphtheria, influenza, malaria and tuberculosis.

## The links between some diseases and geographical features and pathways

The spread of many diseases, like measles, may follow a diffusion model (see Figure 45).

**(a) Expansion diffusion:** the disease spreads from one place but remains concentrated there

**(b) Relocation diffusion,** e.g. some types of flu: the disease moves from place to place

**(c) Contagious diffusion,** e.g. measles: direct contact is needed to pass the disease on

**(d) Hierarchical diffusion, e.g. HIV, SARS:** the disease spreads from one large centre to a number of smaller centres

**Figure 45** Types of spatial diffusion

# Enquiry question 3

# What is the link between health risk and pollution?

Environmental health is a large subset of the health risk problem. There is a range of risks associated with one-off pollution incidents, and longer-term or sustained pollution.

One-off pollution incidents may be moved by air or water along a pathway. They range from low-risk oil spills to higher-risk processed or radioactive substances, e.g. Chernobyl 1986.

Examples of longer-term or sustained pollution include:
- 1980s agricultural nitrates in groundwater linked with 'blue baby syndrome'; in 2009 the EU banned 22 commonly used chemicals in agriculture
- 1970s onwards, melanoma from CFC destruction of protective ozone layer
- climate change — fears of spread of disease
- smoking environments creating carcinogenic risks — WHO aims for a worldwide ban in public places

Knowledge check 51

What types of country, in terms of level of development, appear to be most at risk from severe pollution?

Pollution from land, water or air sources may pose health risks. **Transboundary** effects add complexity, especially from atmospheric pollution hazards like ozone depletion and enhanced global warming. Pollution has escalated globally over the past century and increased in spatial extent with specific **hot spots** or **clusters** often created by point source incidents like Chernobyl, Bhopal and Harbin. Transition economies such as the BRIC group now suffer the most risk. Many older economies have reduced their land and water pollution rates dramatically with deindustrialisation. However, there are now global threats from the health risks associated with climate change.

**Table 25** Checklist for pollution health risk case study

| Causes | Impacts | Management |
|---|---|---|
| Pollution types/sources? | Short term? | Who is involved? |
| Location? | Long term? | International governments? |
| Point/diffuse? | Direct/indirect impacts? | TNC/NGO/Foundation? |
| Incidental/sustained? | Actual threat to humans? | Local pressure group? |
| Atmospheric/hydrosphere/terrestrial? Noise? | Evidence of pollution fatigue — public outcry? | Short-term/long-term management? |
| Caused by economic activity? Primary/secondary/tertiary/quaternary? | Pressure group? | Type of strategy: prevention at source/treatment at sink? |
| Incident or sustained pollution? | Internet site/blog? | Strategy used, why? |
| Where is its location on environmental | | Policies used, why? |
| Kuznet curve? | | Evidence of sustainable management? |

**Research activity**

Research the health risks from enhanced global warming and climate change using your AS notes and references such as the 2004 *Up in Smoke* report on sub-Saharan Africa. Compare the management of chemicals used by EU farmers with those of Green Revolution farmers such as in the Punjab, India, where there may be health consequences of long-term use of DNA.

The link between pollution, economic development and changing health risks may be focused around the **Kuznet** environmental model, based on the Western world's experiences.

**Examiner tip**

In your report, you should include models such as the Kuznet's model if relevant. These could be drawn and, better still, adapted and personalised. Don't spend too long drawing them. Keep them simple and clear.

**Research activity**

Research the Kuznet environmental model (Figure 46) and assess how this model fits rapidly changing areas like Russia, China or India. Does it link with the epidemiological model?

Contrast the management of pollution in the UK with the management in a rapidly developing country such as China. City-based data are readily available on the internet for cities like London, Mexico City and Shanghai.

Look at international policies, such as the latest EU Directives on agricultural chemicals, water quality or electronic waste disposal. On a global level, assess the effectiveness of the UN Basel Convention that restricts transnational trade in toxic waste. Key resources: Blacksmith Institute, Greenpeace, Friends of the Earth.

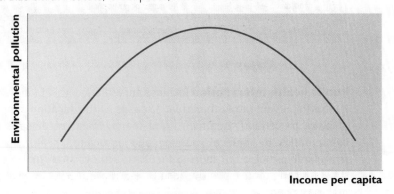

**Figure 46** Kuznet's environmental curve

# Enquiry question 4

# How can the impacts of health risk be managed?

## The socioeconomic and environmental impacts of health risk

The short- and long-term impacts, or burden, of health risk, are now more complex with the so-called 'health divide' becoming an increasing issue for the sustainability of our natural and socioeconomic environment. Health is a major driver of global and local economies, but the costs of healthcare are escalating because of:

- population increase, especially an ageing population
- a rise in poverty and a more vociferous middle class with higher expectations of healthcare
- technology and medical expertise — availability of expensive technology and care in prevention and treatment.
- consumer demand, increased by the media, internet information and demands for more social equity in healthcare
- rising pollution and environmental health risks from workplaces and indirectly from climate change
- global interconnectivity — globalisation of health expectations and faster movement of infectious diseases because of migration and travel. Also media-led panic about diseases such as SARS, avian influenza and swine flu

## How health risk impacts have led to differing management strategies

Some health risks and environments are harder to manage than others (Figure 47). Health risk cannot be tackled on its own since it relies on the complex socioeconomic structures of any society.

| HARDER? | • Longer-term health risks: often chronic diseases, e.g. depression, obesity and diabetes<br>• Also incurable infectious diseases, e.g. HIV/AIDS | • Shorter-term health shocks: mental and physical traumas associated with disasters, plus infectious diseases resulting from breakdown in built environment (loss of water, sewage, housing, food supply)<br>• Longer term relief as seen in 2008 in the Burma floods, and refugee camps in Democratic Republic of Congo | EASIER? |
|---|---|---|---|

**Figure 47** Spectrum of difficulties in managing health risks

**Public health intervention** includes surveillance, vaccination and family planning. It aims to prevent rather than treat diseases, with education a top priority. Healthcare involves prevention, treatment, and management of disease for individuals and communities by medical, nursing and associated health specialists. Prevention is preferable, but often more difficult to achieve than treatment, although simple schemes can produce positive results, e.g. mosquito nets preventing transmission of malaria.

Faster and more publicised types of health risk (such as HIV/AIDS, SARS) get higher priority than, for example, mental illness.

Latest concerns centre on obesity-related diseases, partly due to the globalised **food transition** involving higher fat and sugar intake. By 2015, 2.3 billion adults may be overweight and over 700 million obese (WHO).

Well-established **health systems**, such as in the UK, aim to promote, restore or maintain health. They have an integrated set of facilities and personnel, and have a hierarchy of primary, secondary and tertiary care. They have evolved from informal, small-scale, often family-based systems, into large, often government-run systems. There has also been a rise in healthcare provision by private companies and NGOs, e.g. BUPA, Red Cross.

Growth of international efforts to tackle health risks is linked to:
- increasing scale and issues involved
- globalisation and interconnectivity of world economies, plus political and financial linkages

International **minimum** and **ambient standards** are now common, as are flows of money, personnel and technology across the world. These are funded by international institutions like the United Nations or World Bank, or philanthropic NGOs ranging from Oxfam to the Bill and Melinda Gates Foundation. However, whether policies can be implemented effectively is still up to individual nations and individuals.

## Differences in the effectiveness of health risk management, and the role of sustainability

The largest international health-risk initiative to date is the 2000 UN Millennium Declaration (the Millennium Development Goals or MDGs), adopted by 189 countries, which created a '**roadmap**' for **health sustainability**. Health is linked directly to many of the goals and contributes indirectly to the others (see: **http://go.worldbank. org/RJWW34HS90**). Reduction in diseases like HIV/AIDS, malaria and tuberculosis is expected, but most countries are currently off track, especially sub-Saharan African countries and India (World Bank 2008). This is despite the doubling of health-related aid in 2000–06 from a whole range of players.

### Research activity

There is no doubt that today China presents one of the largest challenges to combat pollution and reduce health risks and is worthy of closer research, especially on how its latest 'Healthy China 2020' plan is likely to succeed. You need to set up criteria to assess effectiveness, such as capital and running costs. Also, involvement of local community and effects on health (use indicators like death rates). Key references: WHO report on disparities in health risk China: **www.who.int/macrohealth/action/CMH_China.pdf** and *New Scientist* October 2008 on China's health system: **http://tinyurl.com/3dn4jun**

The most important single global advisory organisation is WHO, which has projections for future health scenarios, and an overall goal of reaching an 'age of sustaining health' (Figure 48).

### Knowledge check 53

What is meant by the 'food transition' (also called the nutrition transition)?

### Examiner tip

It is important to recognise that different countries have very different healthcare systems, for instance private health insurance in the USA versus the 'free at the point of use' NHS system in the UK.

### Research activity

Contrast the healthcare system of the UK with that of China, or the USA. All are struggling at different scales and in different ways to reduce risks. Find out the range of agencies responsible for healthcare in your own community.

### Knowledge check 54

Why do you think so many of the MDGs are linked, directly or indirectly, to health?

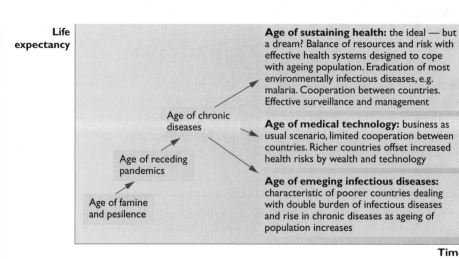

**Figure 48** Health futures: WHO health transition model

## Summary

- Know that there is a wide range of human health risks, some of which are chronic and long term whereas others are short term, such as infectious diseases.

- Research patterns of health risk at a global scale, such as the obesity problem or a named pandemic, as well as national and local patterns in terms of health risks in different locations and demographics.

- Understand the epidemiology model and how health risk changes over time.

- Investigate how health risk affects different groups of people by reducing quality of life and reducing life chances.

- Examine the causes of health risk and recognise that these are complex and related to income levels, pollution, management and the availability of healthcare.

- Explore how named examples of health risks are affected by socioeconomic status and also how geographical features can affect the incidence and spread of diseases, for example transport pathways. You should also be able to relate these risks to models such as the diffusion model.

- Research a range of sources of pollution, both point source and diffuse, and understand how these impact on human health. This should include both one-off pollution incidents as well as more long-term, sustained pollution.

- Examine pollution in a range of countries at different levels of economic development to understand changing risk; this can be related to a critical understanding of the Kuznet curve and other models. The role of pollution management in reducing risks to human health shoud also be considered.

- Consider the impacts of health risk and how these affect people's lives, as well as the burden placed on society, for example the costs of healthcare.

- Examine a range of different ways of managing health risks, ideally in contrasting locations (developed vs developing world) and for both infectious diseases and degenerative conditions.

- The role of different organisations and players — governments, NGOs, IGOs such as the WHO — in managing health risks should be considered through examples and case studies. A study of managing short-term risks (e.g. a pandemic) should be contrasted with the approach to a longer-term risk such as HIV/AIDS; in addition, the question of whether management strategies are sustainable should be considered.

# Consuming the rural landscape: leisure and tourism

## Introduction

**Leisure** is a broad term that traditionally means non-work time. **Tourism** is usually taken to mean the temporary or short-term movement of people to destinations outside places where they normally live and work, and their activities during their stay at these destinations. Figure 49 shows the relationship between leisure, recreation and tourism, and Figure 50 illustrates the range of leisure activities.

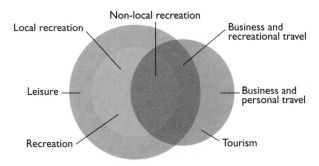

**Figure 49** Relationship between leisure, recreation and tourism

Tourism is economically important — for example, the tourism industry in the UK is worth around £85 billion, more than 4.5% of the UK's GDP (2011 data). There have been three main drivers that have led to the commercialisation (or 'consumption') of many rural landscapes:

- **Economic changes** such as falling farm incomes have meant that new uses must sometimes be found for rural land if local communities are to gain sustainable incomes.
- **Social changes** have left many sectors of people finding that they have greater time available to devote to leisure and tourism activities.
- There has been a **cultural shift** — people's view of the countryside has changed and it has become something to cherish and defend, rather than simply exploit for food and minerals.

**Post-productionism** is the change in farming, moving from maximum food production to more sustainable agriculture and diversification into leisure activities in rural areas. In particular there has been increased emphasis on more active pursuits such as cycling and walking, together with utilisation of the cultural (including food) aspects of the countryside. This is in contrast to activities that were popular 50 years ago such as 'going for a drive' or 'picnicking'.

A more intensive exploitation of the countryside for leisure has not come without costs. There has been increased pressure on often fragile landscapes, together with displacement of local traditions and communities. Conflicts are increasingly common. All these issues require careful analysis and management to find solutions that are economically, socially and environmentally sustainable.

**Knowledge check 55**

How can rural honeypots be regarded as a positive aspect of nature conservation?

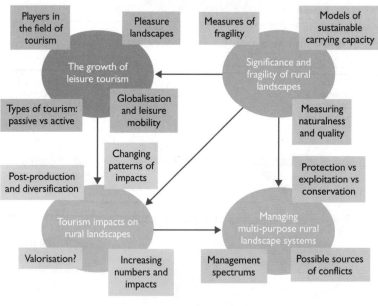

**Figure 51** Topic concept map

**Examiner tip**

Frameworks may be very important for this type of question — justifying a clear focus for selection of locations and/or case studies.

**Figure 52** The synoptic context

Edexcel A2 Geography

Table 26 Checklist for consuming the rural landscape

| Enquiry question 1 What is the relationship between the growth of leisure and tourism and rural landscape use? | Enquiry question 2 What is the significance of some rural landscapes used for leisure and tourism? | Enquiry question 3 What impact does leisure and tourism have on rural landscapes? | Enquiry question 4 How can rural landscapes used for leisure and tourism be managed? |
|---|---|---|---|
| • Active vs passive pursuits<br>• Reasons for changes over time — becoming more diverse<br>• Leisure vs tourism<br>• Different groups and players — their roles and responsibilities<br>• Possible sources of conflicts and between whom<br>• Values and attitudes of different users and beneficiaries | • Landscape values (remote, accessible, wilderness etc.)<br>• Ecological vs physical (utility) values<br>• Notion of fragility and sensitivity<br>• Wilderness concept<br>• Qualitative vs quantitative environmental measures<br>• Sustainable carrying capacity | • Negative impacts of excessive use: trampling, pollution, erosion, disturbance etc.<br>• Positive impacts of tourism: conservation, increased awareness, protection of heritage sites<br>• Changing impacts over time (increasing or decreasing?)<br>• Comparing threats and opportunities in areas with different levels of development | • Arguments for and against management of rural landscapes<br>• Developing a 'conservation spectrum'<br>• Players involved in conservation — choices and conflicts<br>• Evaluating the effectiveness of different approaches and options |
| Key refs used | Key refs used | Key refs used | Key refs used |

# Enquiry question 1
# What is the relationship between the growth of leisure and tourism and rural landscape use?

## The consumption based economy

Wilderness areas are heavily protected against development so leisure and tourism is generally restricted and highly managed. Many locations remain inaccessible and off the beaten track – they are exclusive. Landscape quality is high and often well protected by legislation. Typical activities could include:
• wild camping and trails
• tourist trails
• wildlife spectacles (low key)/tourist gaze, e.g. Antarctica
• engaging with local culture, e.g. Masai village, Kenya
• sustainable ecotourism in rainforests

More accessible rural areas may be dominated by a post-production agricultural landscape, together with agri-environmental schemes. Large numbers of tourists can be accommodated in honey-pot areas. Car parking and visitor facilities may be common. Examples may include:
• organic farms with visitor centre/speciality food shop
• tourist accommodation and holiday lets
• craft centres
• wildlife/'touchy-feely' farms
• paint-balling and quad biking
• woodland trails (e.g. mountain biking)

Highly accessible rural landscapes (typically rural–urban fringe or urban hinter-land) are often battlegrounds between conservationists, planners and developers. Conflicts may occur over both activities and development, especially near to greenbelts or other specially designated areas:
• horse stabling and livery
• golf courses
• craft shops/farm shops
• garden centres/horticulture
• barns for small rural enterprise
• organic food-box schemes
• fish (trout) farming
• dog breeding
• vineyards

**Research activity**

Use Table 26 as an initial stimulus and carry out your own research audit to obtain background sources of information. Make a note of any key references used.

**Figure 53** Different types of leisure and their location

Rural tourism encompasses a wide range of activities. Research these contrasting types of activity:

• passive or 'traditional activities', e.g. fishing, walking, driving
• active or 'new activities', e.g. paint-balling, quad biking
• ecotourism and elite tourism compared to mass tourism

How and why are these activities different and how did they emerge? Where do activities such as holiday cottages and theme parks fit into this picture? Visit Tourism Concern for some introductory ideas: **www. tourismconcern.org. uk/**

One of the most significant changes in the countryside has been the transition from an economy based on food production to one based on consumption (hence 'consuming' in the title of this option). This consumption based economy is diverse and includes things like financial services as well as the more obvious leisure and tourism activities. Figure 53 shows different types of leisure that are associated with particular locations along a spectrum from rural–urban fringe to remote wilderness areas.

An important concept linked to the development and geographical spread of tourism is the **pleasure periphery**. This concept tries to show the spread of tourism, from a point source or single location, to a much more diffuse or global activity:

• 1800: source — close to home/local
• 1900: periphery (1) based in northwest Europe
• 1930: periphery (2) extends to western Mediterranean
• 1950: periphery (3) includes all of the Mediterranean
• 1970: periphery (4) long distance travel becomes more readily available
• 1990: periphery (5) tourists are able access the world's remotest places

The growth of tourism and leisure can be linked to several key drivers:

• increasing family wealth and income
• more spare time for leisure
• increased holiday (paid)
• greater personal mobility
• greater awareness of far-off places through media, e.g. internet, books, magazines
• cheaper air travel
• rising awareness of extreme sports

## The range of landscapes

There is no universally agreed definition of 'rural' but population statistics often form the basis for measurement. For example, in the UK, a settlement of less than 10,000 people is *not* classified as an urban area and by default is rural.

The distance a landscape or activity is from a large centre of population can have a significant impact on its characteristics. Some geographers classify 'wilderness' as being the most extreme rural landscape (see below), while the rural–urban fringe is the most accessible.

**Concept of wilderness:** Wilderness areas are remote parts of the world whose unspoilt characteristics have ecological, scientific and/or cultural and aesthetic value. Wilderness areas can be big (e.g. continental areas of the Americas) or smaller (e.g. corners of rural Britain — such as the granite tors of Dartmoor).

Wilderness landscapes are commonly perceived to be those that have avoided being modified by any significant human impact. However, many landscapes perceived as 'natural' have been modified by people. For example, the UK's 'wild' moors were historically cleared of all trees to make way for sheep farming and grouse shooting. Find out more at The Wild Foundation: **www.wild.org**

**Research activity**

Copy and complete the table below, which looks at the different players involved in rural landscapes and their possible attitudes towards development. With arrows (last column) link together any possible sources of conflicts.

| Leisure and tourism players | Examples | What they do and attitudes to rural landscapes | Possible conflicts (link with arrows) |
|---|---|---|---|
| Governments | | May provide funding/legislation to promote diversification, development etc. Likely to have social or economic benefits | |
| Intergovernmental agencies | | | |
| NGOs and pressure groups | Greenpeace, Wildlife Trusts | | |
| Local/regional authorities | Natural England | May help in the promotion of rural landscapes to diversify economy, e.g. 'surfing-tourism' in southwest England. Can potentially create hot spots and therefore traffic issues, pollution, congestion etc. | |
| Communities | | | |
| Farmers and local businesses | | | |
| Individuals | | | |
| | | | |

# Enquiry question 2

# What is the significance of some rural landscapes used for leisure and tourism?

The value of rural landscapes cannot be easily defined in terms of monetary worth. Historical and cultural resources are intrinsic components of the landscape, but difficult to price. The ecological and physical value of landscape are particularly significant and again it is problematic to assign a monetary value to these. So landscape value should be regarded as having three components:

1 **Ecological and physical value** — the condition, attributes and content/resources of a landscape.

2 **Socio-cultural value** — how it contributes to physical and mental health, learning etc.

3 **Economic value** — as quantified by direct market value, e.g. land price if it were sold.

Many rural landscapes that have high ecological value may also be fragile. Fragility refers to how stable a landscape is, or how easily it can be damaged. To assess the sensitivity and fragility of a landscape's character decisions have to be made with regard to:

- liability/potential loss of a landscape's key attributes and characteristics
- whether or not the landscape could be restored (this may be particularly important for ancient landscapes, e.g. woodlands pre-1600 or old monuments)
- whether important visual or aesthetic aspects of its character would be likely to change

What is meant by fragility and sensitivity in the context of landscape management?

Consideration should be given to the introduction of new elements that may have a significant impact on landscape character. This could include new buildings and infrastructure such as road, rail, airports, power lines, data cables and pipelines, as well as invasive non-native species of plants and animals. Figure 54 gives a more detailed example of how landscape fragility and sensitivity can be determined.

**Landscape character sensitivity**
*Based on judgements about sensitivity of:*

**Natural factors**
Vegetation types
Tree cover type/pattern
Extent and pattern of semi-natural habitat

**Cultural factors**
Land use
Settlement pattern
Field boundaries
Enclosure pattern
Time depth/historical significance

Overall landscape sensitivity and fragility =

**Landscape quality/condition**
Intactness
Representation of typical character
State of repair of individual elements

**Aesthetic factors**
Scale
Enclosure
Diversity
Texture
Pattern
Colour
Form/line
Balance
Movement

+

**Visual sensitivity**
**General visibility**
Landform influences
Tree and woodland cover

**Population**
Numbers and types of residents
Numbers and types of visitors

**Mitigation potential**
Scope for mitigating potential visual impacts

**Figure 54** Factors to consider when judging overall landscape sensitivity

**Research activity**

Carrying capacity is an important idea. Find out more about the different types including (1) physical carrying capacity, (2) ecological carrying capacity, (3) social/psychological carrying capacity (see Figure 55) and (4) economic carrying capacity. These may be particularly important when set against a background of access capacity, commercial capacity, construction capacity, service capacity and transport capacity. For active tourism see:
**www.active-tourism.com/DefinitionActTour.html**

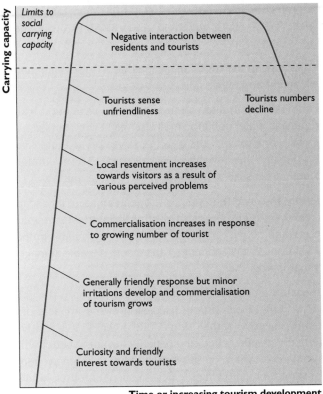

**Figure 55** Social carrying capacity

Quantitative and qualitative methods can be used to measure landscape quality and ecological value:

- **Quantitative** means numerical data — for example a species biodiversity count or a species frequency, e.g. using a transect, or volume of timber in an area or an actual land value price. An advantage of this method is that data can be readily processed using statistical approaches. Numerical values can also be used to provide an array of data for GIS visual processing systems.
- **Qualitative** means subjective or non-numerical data. This may include the overall 'look' or 'feel' of the landscape and could be assessed using an environmental quality index, field sketches or photographs. However, these data can be subjective and based on interpretations made by researchers as they explore their surrounding environment. Often qualitative judgements are considered unreliable.

**Knowledge check 57**

What is GIS and how can it be useful to assess landscape quality?

# Enquiry question 3

# What impact does leisure and tourism have on rural landscapes?

## Positive and negative impacts

Impacts of people on the rural landscape can be positive as well as negative. Over the last 200 years, legislation has been passed to grant animals, plants, ancient monuments etc. protection from harm or exploitation. Important benchmarks include the formation of the RSPCA (1824) and RSPB (1899), IUCN (1973) and the establishment of World Heritage sites (1972).

An example of a change in the rural landscape that has provided much debate is the continued development and expansion of golf courses.

Popular (especially honeypot) sites often experience vegetation trampling close to footpaths. Air, land and water pollution may also blight rural landscapes, even remote wilderness regions. For example, the beaches of Antarctica and Alaska are strewn with the plastic flotsam that has been lost overboard from commercial shipping. Areas that are more accessible will usually receive more tourists.

While it is generally assumed that more visitors mean more landscape impacts, this is not always the case. High impact sites often have greater regulation to control visitor movements and activity patterns. In contrast, a small number of visitors in unregulated or wilderness environments can do considerable damage by felling trees for firewood and dropping litter.

Emphasis on negative environmental impacts can obscure the fact that tourists can bring benefits. Any research and discussion of the impacts of tourism needs to be balanced and objective, rather than a 'tale of woes'. It is relatively easy to find examples of tourism revenue generating benefits including better wildlife protection and conservation, restoration of historic buildings, footpath repairs and even improvement of sanitation following the development of tourist hotels.

Use the table below as a structure for assessing the positive and negative effects of leisure and tourism. You should use examples to support your ideas. How and why do the threats or opportunities associated with rural tourism vary according to the level of development of the country/region?

| | Positives | Negatives |
|---|---|---|
| Social | | |
| Environmental/ physical | Restoration of historic buildings, upgrading of resorts, improvement of polluted environments... | Soil erosion (especially following deforestation), depletion of local resources, damaging construction traffic, visual pollution... |
| Economic | | |

## Commodification of the rural landscape

The promotion of local produce and traditional or specialised food products can be an important component of the valorisation or commodification of the rural landscape. Here there is emphasis on direct sales to consumers by farmers and local producers. Moreover, the sites of local food production, such as farms, dairies, vineyards and breweries are additionally marketed as tourist attractions, creating a second income stream and reason to visit.

Another way in which a rural area may market or rebrand itself is through the use of film and new media. Table 27 gives examples of a range of films that have helped promote the rural landscape.

**Table 27** Films and television programmes that have promoted the rural landscape

**UK locations**

| | |
|---|---|
| *Harry Potter* series (2001–2011) | Northumberland (Alnwick) |
| *Pride and Prejudice* (2005) | The Peak District |
| *The Wicker Man* (1973) | Dumfries and Galloway |

**International rural locations**

| | |
|---|---|
| *Passage to India* (1984) | Marabar Caves, nr Chandrapore, rural India |
| *Narnia* series (2005–present) | Poland (Table Mountains) and New Zealand |
| *Lord of the Rings* series (2001–03) | New Zealand (Kaitoke Regional Park) |
| *The Beach* (2000) | Thailand (Phi Phi Islands) |
| *Star Wars* (1977) | Tunisia (Chott el Djerid) |
| *Brokeback Mountain* (2005) | Canada (Alberta) |

Television also establishes honeypots for rural tourism. *Balamory* is a children's show that began filming in 2002 on the Scottish island of Mull, which subsequently experienced an influx of tiny tourists (accompanied by their parents). *Heartbeat* (Yorkshire) — rural set locations have become popular with tourists.

# Enquiry question 4

# How can rural landscapes used for leisure and tourism be managed?

Many different groups of people and organisations (players) are important in controlling the relationship between rural landscapes and varying forms of tourism and leisure. Some of these organisations hold direct power through access to funding, while others exercise a broader influencing or advisory role.

Rural areas are sometimes viewed by politicians as 'problem regions'. This is because they are typically poorer (more limited employment opportunities) and less accessible than their urban counterparts.

When considering the management of rural areas there are a series of questions that must be evaluated:

- What rights should indigenous people/landowners hold over their land and its ecosystems?
- Should endemic/localised species (including those that are dangerous) be restored to rural landscapes?
- To what extent should rural landscapes be restored, wherever possible, to an 'authentic' or original state?

Conservationists and land managers need to be clear about what they are trying to achieve and what outcome they require (Figure 56). In the UK, sites are often micro-managed or preserved for a particular habitat type, e.g. rich grassland, ancient woodland or heather moorland.

There is a new EU mandate that establishes a duty for all member states to re-establish their native flora and fauna. In the UK this will mean the re-introduction of animals such as beavers, wolves, red kites and sea eagles.

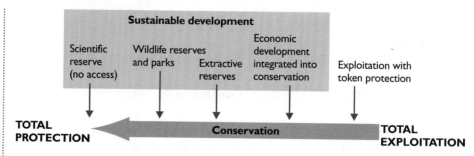

**Figure 56** Spectrum of ways in which rural landscapes can be managed

There are various approaches to management. Table 28 provides some examples that can be applied in rural areas.

**Table 28** Approaches to management of rural landscapes

| Direct | Indirect |
|---|---|
| Often used in most fragile areas or in a potentially dangerous situation, e.g. waterfall, crumbling ruin. <br>• Most time consuming and expensive <br>• May need to start with this in short term to protect, and then move to more indirect means as education kicks in <br>• Regulations that may entail enforcement, restricting activities or rationing use | Usually more successful in remoter locations and cheaper. <br>• Seeks to affect behaviour through education, information and persuasion <br>• Visitors can be informed about the impacts connected with a certain activity, or given information that encourages the use of certain areas over threatened areas <br>• Physical alterations, such as the redirection of a trail to a more resilient area of a forest, that influence the movement of visitors |
| **Hard** | **Soft** |
| Paths, fences, vegetation clearance, reseeding... | Land use zoning, litter bins, interpretation signs and centres, nature trails... |

**Knowledge check 61**

What is the history and relevance of the National Parks and Access to the Countryside Act?

**Research activity**

Research how National Parks are managed; identify the key players and stakeholders, and consider what are the possible sources of conflicts. Key reference: the UK National Parks website: **www.nationalparks.gov.uk/**

Summary

- You should research why there is increasing interest in the importance of rural areas as spaces for recreation and tourism, and how such areas can have a wide variety of character profiles (found along the rural–urban continuum).

- What are rural areas and why is it not appropriate to provide a 'one size fits all' definition that can be applied in different countries?

- Explore how increasing urbanisation on a world scale (but especially in emerging economies) is putting a range of pressures on rural resources, including issues of abandonment and neglect.

- Recognise how rural landscapes have a number of sometimes competing uses — they are multipurpose. Explore their physical, ecological and economic values.

- Investigate and contrast different importances of rural landscapes in terms of ecology and fragility, for example compare the Lake District with the Masai Mara.

- You should research examples and case studies of rural landscapes that have different carrying capacities and levels of resilience. Understand the notion of 'thresholds'.

- Investigate the process of Environmental Impact Assessment as a tool for deciding what and where to protect and how it has been used in different places as part of the planning process. What are the advantages and disadvantages of such an approach and are there any more suitable alternatives?

- Research how political stability, careful land management and new technologies might help reduce damage in ecologically sensitive and fragile rural environments. What role does the local community have in making places more resilient?

- Contrast different types of technology that could be used to both manage and predict damage to rural areas, especially the role of GIS and new land management tools. How can 'apps' be used to help visitors explore locations in a more meaningful and managed way?

- Consider the role of players such as governments, IGOs and NGOs in managing rural landscapes, both in the short, medium and longer term.

- Investigate the spectrum of approaches to rural land management, evaluating the advantages and disadvantages of each and supporting with relevant case study information.

# Questions & Answers

In this section there are six questions and responses — one for each option. You should browse through all the responses even though you will only study one option — you will find many transferable tips for the report.

The questions and research focus are similar in style to those given in the real examination. Remember that each question is worth a maximum of 70 marks.

Sample student answers are provided for each question. They illustrate a mix of A-grade and C-grade responses. The answers provided are not complete reports but are extracts from typical responses, e.g. the introduction, main body or the conclusion. Each answer is followed by examiner's comments (indicated by the **e** icon), which explain where credit is due and where weaknesses or errors occur.

Before each answer is a section that helps 'unpick' the research focus. This highlights the key areas in which the options will be assessed. It includes suggested issues for you to cover, and help on possible case studies.

In the examination, answers will be marked using the generic mark scheme summarised below.

| | | Marks |
|---|---|---|
| **D** | Introducing, defining, focus | 10 |
| **R** | Research and methodology | 15 |
| **A** | Analysis, application and understanding | 20 |
| **C** | Conclusions and evaluation | 15 |
| **Q** | Quality of written expression and sourcing | 10 |
| **TOTAL** | | 70 |

# Exam guidance

## How to use the mark scheme

Part of the mark scheme is shown below. It is **generic** because it applies to any exam question. The mark scheme sections are known as DRACQ:

D = Discuss/define

R = Research and methodology

A = Application and understanding

C = Conclusions and evaluation

Q = Quality of written communication

Only the top level for each of the DRACQ criteria is reproduced here. You should get a full mark scheme from the Edexcel website **www.edexcel.com** to see how to maximise your learning and skills in the final exam.

Some students find it useful to remind themselves of the DRACQ on their planning page. You will see that writing a full introduction and a thorough conclusion with sub-evaluations in the report will gain you 25 of the available 70 marks. The application criteria are the most critical — as always in exams it is not so much what you know but how you use it.

Crucial to gaining high marks is to practise writing reports: first as research tasks, probably longer than you would be able to write in the exam, and then under timed conditions.

**Table C** The top-level mark bands from the generic mark scheme

| | | Introducing, defining and focusing on the question — 10 marks |
|---|---|---|
| **D** | 9–10 | • Clear reference to title — develops a focus<br>• Indication of framework, by concepts and/or case studies<br>• Accurate definitions of key terms |
| | | **Researching and methodology — 15 marks** |
| **R** | 12–15 | • Wide range of relevant case studies used (by scale and/or location)<br>• Relevant concepts, and/or theories used<br>• Factual, topical evidence<br>• Indication of methodology, i.e. how evidence was sampled/selected |
| | | **Analysis, application and understanding — 20 marks** |
| **A** | 17–20 | • All research applied directly to question set<br>• High conceptual understanding and cogent argument<br>• Appreciation of different values/perspectives about the question<br>• Any maps/diagrams are used to support answer |
| | | **Conclusions and evaluation — 15 marks** |
| **C** | 12–15 | • Clearly stated<br>• Thorough recall of content/case studies used in essay<br>• Ongoing evaluation throughout report<br>• Understands the complexity of the question |
| | | **Quality of written communication and sourcing — 10 marks** |
| **Q** | 9–10 | • Coherent structure and sequencing with obvious report style sub-sections<br>• Excellent standards of spelling and punctuation<br>• Geographical vocabulary used correctly<br>• Diagrams/maps, if used, incorporated into text and support argument<br>• Referenced/acknowledged material: obvious evidencing/sourcing from wide range of sources (texts, journals, internet, DVDs etc.) |

## Getting the best out of the pre-release research focus

The following format will be used in the pre-release research focus:
- **Explore...** This will guide you to the enquiry question/s and sub-questions chosen by the principal examiner. It refers to the background concepts, processes, theories and models involved, where relevant.

- **Research...** This will hint at the type of supporting evidence needed, for example a range of areas/types/strategies, and means a focus on the geographical places, case studies and examples illustrating these.

Here is an example from Option 5 Pollution and human health at risk.

### Pre-release focus

- **Explore** the varying links between the cause and spread of health risk and geographical features.
- **Research** a range of types of health risk and locations.

### Unpicking the research focus

You could make a table and take a systematic approach, as shown in Table D.

**Table D**

| Pose a question from the research focus | Which enquiry questions? | Details | Actual case study you could use/need more research on |
|---|---|---|---|
| What are health risks? Range needed? | EQ1: Types, patterns, globally, over time | Range: chronic, infectious, traumas | Range, e.g. by scales: small–global<br>Economic pathway: MEDC–NIC–LEDC |
| What are the causes? What are the geographical features? Spread? | EQ2: Complex causes, relationship to socioeconomic status<br>EQ2: Links between some diseases and geographical features at variety of scales. Development and spread<br>Model: diffusion<br>EQ3: Pollution<br>Also a bit of EQ4 on management, which may reduce worst effects/increase problem if mismanaged | **Human**<br>• Infrastructure, transport<br>• Population concentrations, urban focus?<br>• Lifestyle choices, economic development<br>• Poverty<br>• Type of industry/employment and role of pollution, e.g. radiation, chemical spills | • Swine flu 2009 — air travel, intensive pig farming, some mismanagement by WHO and Mexico authorities at start of pandemic; use relocation diffusion model<br>• Chernobyl, nuclear-radiation<br>• Lifestyle feature: obesity and fat camps of UK, USA, China<br>• Cholera: political corruption and famine vulnerability — Zimbabwe 2008 |
| | | **Physical**<br>• Role of water especially rivers<br>• Role of climate change and spread of some risks | • Cholera refugee camps Darfur 2006<br>• Malaria Kenya twenty-first century |
| Varying links? | Means some links are strong, some weaker | Strong — possibly direct<br>Weaker | Shanty town conditions, industrial pollution<br>Political and social reasons, e.g. HIV/AIDs? |

## Tackling exam questions: command words and key words

Every examination question in this unit will have the following components:

- **Command word/s** — e.g. evaluate, discuss
- **Key words** — for focus of the discussion, e.g. challenge, relationship, spatially
- **Geographical focus** — a focus for your discussion of the question, e.g. food security issues, responses to tectonic hazards

Table E gives a list of common command words used in exam questions.

**Table E** Common command words

| Command word | Meaning |
|---|---|
| Assess | Weigh up both sides of an issue/solution and come to a conclusion |
| Consider | Describe and explain a range of different views on a subject; explore a range of positions |
| Discuss | Give both sides of an argument (for and against), and come to a conclusion |
| Evaluate | Weigh up several options or arguments and come to a conclusion about their importance/success |
| Examine | Investigate in detail; offer evidence for and against |
| Explain | Provide a detailed set of reasons why something is like it is |
| To what extent | Say 'how far' you agree with a statement or option by examining its advantages and disadvantages |

Your report needs to show that you can both develop an argument and sustain it. Plain descriptions of case studies will not win you many marks at A2, but using an evaluative style will get you into the highest levels. An evaluative answer explores several sides of an argument, and shows that geography is complex and there is no 'one' solution or viewpoint. Certain words can be used to develop an evaluative style of writing, such as: *however, but, although, nevertheless, on the other hand, conversely, despite, whereas*. These show you are considering several sides of an argument.

Key words are instructions that tell you what to write your answer about. They provide you with the *focus* of the question. Some common key words are given in Table F.

**Table F** Common key words

| Key word | Meaning |
|---|---|
| Action | Strategies and methods that might be used to manage a problem |
| Causes | The reasons why something happens |
| Challenges | Difficult, large-scale problems that require solutions |
| Characteristics | The key features of something, often split into: physical/environmental and social/economic in geography |
| Concerns | Aspects of an issue that are worrying |
| Conflicts | The issue over which two or more groups are arguing |
| Consequences | The results of a change or process; they can be positive or negative |
| Distribution | The geographical pattern, most often on a map |
| Effects | The results of a process occurring (comes after causes) |
| Factors | The underlying causes of a problem or process |
| Futures | The range of pathways humans could choose to take, e.g. sustainable or business-as-usual |
| Impacts | The results of a process or change on people and the environment; can be positive and negative |
| Implications | The likely consequences/impacts of a change |
| Issues | Concerns; problems that are worrying |
| Management | Using policies and strategies to minimise or reduce problems |
| Pattern(s) | The distribution of something; where things are; most often on a map |
| Places | Areas of the world, of varying scales |
| Player | An individual, group or organisation involved in an issue (stakeholder) |
| Problems | Issues that worry people; the negative results of a process or change |

| Key word | Meaning |
|---|---|
| Processes | Sequences of events that cause change to take place |
| Relationships | Usually used to mean the links between a cause and its effects |
| Responses | Ways in which people might react to a problem or change |
| Role | The functions or activities of an individual, group or organisation |
| Scale(s) | The size of a feature — global, regional, national, local |
| Strategy | A method used to manage a problem or issue |
| Threats | The causes of negative impacts |
| Variation | How far something differs from the norm or the average; include anomalies (differences from the norm) |

## Tackling exam questions: planning your answer

There are many ways you could approach planning your answer:
- a basic bullet list or a spider diagram
- a spectrum diagram: where you consider two sides of a question: large scale versus small scale, strengths versus weaknesses, things I agree/disagree with etc.
- DRACQ (see above): plan you answer according to the mark scheme
- BUG — **B**ox the command word(s), **U**nderline the key words, **G**lance back to ensure you understand the whole question and choose case studies/examples
- Note down the timings for your answer, e.g. 5 minutes to plan answer, 10 minutes to write introduction, 50–60 minutes for main section of report discussing title and sub-conclusions, 15 minutes for final conclusion and final check

## Tackling exam questions: your introduction

- Identify the command words and respond to them — not to what you did in a mock or practice question at school/college.
- Introduce by focused discussion not by a list of definitions — do not use a pre-learnt introduction.
- Introduce your case studies and a model, if relevant, to answer the question effectively. Avoid too much detail at this stage.
- Start using the correct geographical terminology.
- Start arguing a case if that is what the question asks, but do not give away your final conclusion. You may have changed your mind after looking at your case studies.

## Tackling exam questions: researching and methodology

- Be selective about which research (in terms of examples, case studies, models and concepts) to include and which to leave out; you need to filter your knowledge and include only that which is most relevant.
- Make sure the bulk of the material you use is topical and up to date; if you use old or historical material, you will need to justify its inclusion.
- Try to include some theories and/or models (which can be drawn as diagrams) as well as content, facts and figures.
- You should include methodology that outlines the research material you have used. This could be in the form of a paragraph or table. You should comment on the

reliability of the sources you have used, how up to date they are, the range of material you have used, its bias or questions over reliability.

## Tackling exam questions: your analysis section

- Examiners like diagrams that support/extend knowledge. Make sure any diagram is relevant to the title.
- Maps just showing location are a waste of time: the examiner will know where something is.
- Link the theoretical concepts and models to reality/case studies.
- Try to link up your paragraphs — see the signposting and linkage terms above.
- Remember to link each case study back to the precise title. Try to argue a case — the titles are designed to allow you to do this.

## Tackling exam questions: conclusions and ongoing evaluation

Remember that all-important **conclusion**. It summarises main points/arguments and should have been built up progressively in the report as sub-evaluations.

- You should include summative statements or sub-evaluations within your report, at the end of each of the main sections.
- These should summarise your argument so far, and pick out the key messages from your case studies and examples.
- Ideally, refer back to the title question as part of this ongoing evaluation.

In your main conclusion at the end of your report, you should do the following:

- Summarise the key points of your argument, as a number of general statements that you have proved to be correct.
- If you used a model or concept in the introduction, refer to it again in the conclusion.
- Briefly review the main evidence you have presented.
- Return, directly, to the title question using words and phrases from the title as part of your answer.

Importantly, the conclusion is not a place to introduce new ideas and examples.

## Tackling exam questions: quality of written communication and sourcing

- You must provide references for your main sources of information such as books, websites, DVDs or articles. These can be incorporated into the main text (in brackets) and also should be mentioned in your methodology.
- Your report should be structured, using sub-headings (it is not an essay).
- You should use geographical terminology accurately throughout.
- Your report needs to be logically organised and structured.
- If you use diagrams, give these titles and figure numbers and refer to them in the text.
- Your spelling, punctuation and grammar are taken into account.

# Option 1 Tectonic activity and hazards

## Question

**Discuss the relationship between the nature of tectonic hazards and human responses to them.**

## Pre-release research focus

- **Explore the causes of different types of tectonic hazard and the spectrum of responses to them.**
- **Research a range of examples of different hazard events and the reasons why responses to them differ.**

## Unpicking the research focus

🅔 1 This question covers several enquiry questions from the specification:
- Enquiry question 1: **causes** of hazards and **different events** — suggest specific examples of earthquakes, volcanoes and tsunamis and their nature, i.e. general characteristics. The event profile may help here.
- Enquiry question 4: **responses** — the word **spectrum** means across varying responses (do nothing, adjust, or leave).
- Response will also partly depend on **impact**, so enquiry questions 2 and 3 on impacts must be included too.

2 To narrow down revision, it may help to develop a grid like Table 1 to help you think through the various models and case studies you could choose.

**Table 1** Models and case studies

| Type of event | Do nothing | Adjust | | | Leave permanently or temporarily |
| | | **Modify event** | **Modify vulnerability** | **Modify loss burden** | |
|---|---|---|---|---|---|
| Earthquake | | Experimental in oil fields, Colorado | Aseismic buildings & retrofitting high & low tech. Prediction | Insurance, e.g. California Aid, e.g. Sichuan 2008 | |
| Volcano | Nevada del Ruiz 1985 | Mt Etna 1992 Operation Volcano Buster lava blasting | Monitoring e.g. Ruapehu, N Zealand Zoning of land uses, exclusion zones, e.g. Montserrat & Mt Rainier lahars | IDNDR & Hygo Framework of UN global blueprint for disaster risk reduction 2005+ | Temporary evacuation: Mt Merapi 2006 Mt Chaiten 2008 Long-term: residence flanks of Vesuvius — holiday homes only |
| Tsunami | | | Protect & replant mangroves e.g. Malaysia post 2004 Knowledge e.g. retreating sea Warning systems, e.g. NE Atlantic & Pacific | Aid to, e.g. SE Asia | |

**3** Create some diagrams to identify reasons for different responses (e.g. Figure 1).

**Figure 1** Key factors affecting tectonic hazard response

**4** Check out some easily drawn diagrams to support your report: as there is a focus on the cause of a hazard, you should be able to draw simple cross-section diagrams as well as simple models, for example Figure 2.

**Figure 2** Hazard response and management cycle

**5** Practise quick definitions with sources, for example:
From the UNISDR website:
- A tectonic hazard is a potentially damaging geophysical event…that may cause the loss of life or injury, property damage, social and economic disruption or environmental degradation.
- A disaster is a function of the risk process. It results from the combination of hazards, conditions of vulnerability and insufficient capacity or measures to reduce the potential negative consequences of risk.
For this option we will look at two example introductions.

# Student answers

## Plan

Intro: definitions, e.g. hazard

Middle bit: examples — Mt Merapi, Asian tsunami 2004, earthquake Loma Prieta and Sichuan

Conclusion: sum up

## Introduction

A hazard can be defined as an event or process that has the potential or causes damage to people and their settlements. Natural hazards are physical processes that cause this damage, and can be tectonic, geomorphic or climatic. People who live in areas that experience natural hazards are at risk (vulnerable). The realisation of a natural hazard is a disaster.

The causes of some hazards can be prevented, although this is not usual. Other hazards can be predicted. Some hazards cannot be predicted or prevented so the only management option is to try to mitigate their effects.

In general hazards can be managed in one of three ways as shown the diagram below.

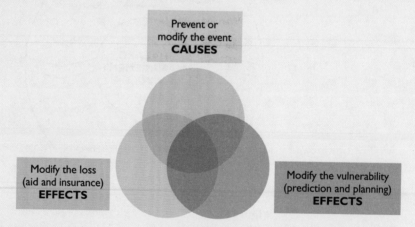

Hazard responses

The diagram above will be referred to throughout this report. Developed countries (such as the USA and Japan) are wealthy, have secure housing, emergency services, media systems and planning bodies. Less developed countries such as India and Indonesia lack this wealth. Levels of social development (literacy, healthcare) are often lower in the developing world and this creates population vulnerable to hazards. This report will focus on the contrast between developed and developing countries in terms of mitigating the impacts of hazards.

ℯ This introduction is likely to be awarded **6/10 marks**. It is a poor plan; there is no real dissection of the question, just a vague list of examples.

There is some framework based on concepts, but no mention of case studies to be used. The focus is likely to be narrow in the rest of the report because the student plans to have a simplistic split between developed and developing economies. This is an outdated concept and only one of the factors affecting response, although wealth may be important as an underlying factor in response.

Definitions of key terms are incomplete — for example, tectonic hazards are not specifically introduced.

### Student B

## Plan
Relationship — graph

**Nature** = type, scale, frequency. **Other factors** = vulnerability: do risk equation **Response** = players, people.
**Defs** = hazard UN & model

Concl — return c/s, complexity, role UNISDR in response
Vocab: aseismic, low/high tech, Benioff, mitigation, salience, pyroclastic

## Introduction
A tectonic hazard is, according to the UNISDR website, a 'potentially damaging geophysical event...that may cause the loss of life or injury, property damage, social and economic disruption or environmental degradation.'

Tectonic hazards are caused by physical processes on the Earth's surface or within the Earth and their very nature in scale, occurrence, frequency and lack of prediction means they are unlikely to be prevented easily. The main focus of response is to reduce vulnerability including *mitigation*. This is shown in the table below.

Tectonic hazards

| Prevented | Predicted | Neither prevented nor predicted |
|---|---|---|
| Lava flows — defences can stop some flows<br><br>*Lahars* — channels can divert and contain them | Volcanoes — can be monitored and eruption time estimated, often with some accuracy<br><br>Earthquakes are more difficult<br><br>Tsunamis, as a *secondary hazard* to an earthquake, may be monitored and since they travel long distances, warnings may be given | Earthquakes — cannot currently be predicted in terms of time, although certain zones are known to be more hazardous, e.g. *subduction and transform faults*<br><br>*Pyroclastic flows* — locations very hard to predict |

A disaster is a function of the risk process. It results from the combination of hazards, conditions of vulnerability and insufficient capacity or measures to reduce the potential negative consequences of risk. The *risk equation* is critical to the sort of relationship that exists between the hazard and response:

**hazard = (risk × exposure × vulnerability (physical, social, economic) × response)**

Responses have shifted since the end of the twentieth century from developing humanitarian response and repair post-disaster, such as in the Mexican City earthquake of 1985, to more mitigation and reduction of vulnerability. This is led by the UN's decade for disaster reduction and 2005 *Hyogo Framework*. This influential *player* in responses is partly a response to the inequalities in ability to reduce vulnerability, especially among less economically developed countries. Wealthier economies (such as the USA and Japan) have more *technological fixes* such as aseismic buildings, emergency services, media systems, literacy and effective civil and planning bodies. The responses of rebuilding and rehabilitation happen everywhere, but funding is usually better in the developed world where insurance exists.

In general, hazards can be managed in one of three ways as shown in the diagram below.

Spectrum of response

The diagram above will be referred to throughout this report with contrasting case studies of well-documented earthquakes like Izmit, Turkey (1999) as well as more recent tectonic events in Sichuan, China (2008) and L'Aquila, Italy (2009) where both a developed economy and a *transition economy* responded quickly post-event. The tsunami in the Indian Ocean (December 2004) will be used to show how the scale of the event may dictate the response, both by those directly affected and by international agencies, and the setting up of the TWS in the Pacific Ocean tries to address a need for reduction in vulnerability by giving information pre-event. Lastly volcanoes, possibly easier to predict and respond to, will be evaluated by using the evacuations of Mt Chaiten (2008) and land-use zoning of Vesuvius compared with the *do nothing approach* of Nevada del Ruiz (1985). Hazard *salience* may reduce response, however, as shown in the complacency in earthquake-prone California where many have forgotten the 'big one' predicted.

@ This is likely to be awarded **10/10 marks**. It has a useful plan and relates to the question. It:
• develops a focus
• includes an indication of a framework, both by concepts and case studies
• gives accurate definitions of key terms, which are referenced from reputable research source. This will help the marks for researching generally

It has already started to accumulate marks for an appropriate report style with good terminology (words in italics).

## Option 2 **Cold environments: landscapes and change**

### Question
**Cold environments are increasingly coming under pressure from conflicting demands. Discuss.**

# Pre-release research focus
- **Explore the challenges that glacial and periglacial environments are increasingly facing.**
- **Research a range of cold environments with differing uses and how effective their management is.**

## Unpicking the research focus
(e) Start by sorting out the **uses** of cold environments, for example with a simple spider diagram.

**Figure 3** Uses of cold environments

Now sort out what **challenges** there are at present, in the past and in the future. The words 'increasingly facing' suggest a temporal element is involved, so you will need to find out whether the challenges have increased, decreased or changed in type over time.

A simple table such as Table 2 may help — add examples as you go.

**Table 2** Challenges: cold environments

|  | Glacial | Periglacial |
|---|---|---|
| **Past** | Past glaciations: challenge of extreme climate conditions, people had to adapt or die, e.g. Inuit N America<br><br>Marginal agriculture, low productivity, e.g. Switzerland, Lake District<br><br>Use for power — HEP<br><br>Antarctic Treaty | Alaskan conflict gas/oil production vs conservation<br><br>Trampling, e.g. Cairngorms |
| **Now, especially climate change** | Melting, recession — Arctic, Antarctica, Himalayas, Rwenzori Mountains E Africa<br><br>Post productive landscapes, increasing role of tourism — Lake District<br><br>More hazards — avalanches<br><br>Increased concern over loss of fragile ecosystems | Increasing melting, thermokarst<br><br>Tar sand exploitation, Nenets of Russia, Trans Alaska Pipeline, Arctic National Wildlife Refuge<br><br>More conservation: Cairngorms National Park |

| | Glacial | Periglacial |
|---|---|---|
| **Future** | Increasing tourism, yet less snow — climate change — artificial snow, new activities, e.g. extreme sports, mountain biking<br><br>Increasing demands on Antarctica — World Park? | Continued |

You need to look at effectiveness of management in reducing challenges. Set up some criteria, e.g. cost, longevity, involvement of local/indigenous people — even sustainability.

Ensure you are clear on the players involved: individuals to international organisations, governments to businesses, IGOs and NGOs.

Create a list of useful vocabulary: relict landform, thermokarst, indigenous, dynamic equilibrium etc.

## Plan

**BUG** the question:

    **B**ox the command word/s

    **U**nderline the key words

    **G**lance back to ensure you understand the whole question and choose case studies/examples

**Cold environments are <u>increasingly</u> coming under <u>pressure from conflicting demands</u>. Discuss.**

The focus is on pressure and conflicting demands (remember the spider diagram from the pre-release stage), so sort out which case studies will be best to show a range of cold environments and a range of demands — you could do a quick table (Table 3).

**Table 3** Cold environments, examples of case studies

| Argument | Glacial | Periglacial |
|---|---|---|
| Yes — more pressure | Switzerland from HEP, tourism, changing climate | N Slope Alaska — resources — conservation |
| No — less pressure because of management | Antarctica so far — resources versus conservation | Cairngorms National Park<br><br>Tourism — conservation |

## Structure

Always start with the main theme of the question — in this case *more* pressure, and then move to the opposing argument, or anomalies to the main trend. Here it is the concept of less pressure because of either reduced demands or mitigation from effective management.

You may be able to structure the report in different ways. Here are some different methods:

• by case study: safe, often pedantic and repetitive

• by type of area: glacial, periglacial, low and high latitudes

• by amount of pressure: high to low to reduced. This is more complex but may get higher marks

• by type of demands and conflicts: e.g. tourism versus conservation and then resource extraction and conservation

• by timescale: past, present, future

The following examples of student work show a conclusion to this question. Read them both and consider the examiner's comments.

# Student answers

## Student A

### Section 3: Final conclusion

The examples used in my report show that cold environments vary greatly in their character, and that relict glacial areas like the Lake District as well as the larger scale Antarctic glacial landscape are experiencing rising pressures. These pressures are often from conflicting demands such as more tourists, or big companies looking for resources. They can also be just from one demand, for example big HEP projects in Canada or the Himalayas. Many people want to keep cold environments as pristine as possible and stop wildlife and ecosystems being disrupted. There is the growing pressure of environmental change because of climate changes. Even periglacial areas are being affected by this, such as in the thermokarst areas of Alaska around Anchorage, where increasing temperatures are causing problems in permafrost cycles and collapsing building structures.

**e** This conclusion does return to some of the concepts and examples used in the report, but it does not return explicitly to all the main ones used and is rather short. At the end it drifts off into general pressures rather than those from conflicting demands on cold environments, and even adds a new case study to the conclusion (thermokarst in Alaska) without connecting it back to the main question.

This conclusion would be awarded **8/15 marks** because it has 'meaningful comments', but only 'selective recall of content' of the report. This response also had some good ongoing evaluation (not seen in this extract).

## Student B

### Section 3: Final conclusion

In conclusion, it can be seen that all cold environments experience pressures from human activity but varying in scale and degree of directness. Some of these pressures are direct, such as in the present largest global dam construction boom of the Himalayas for HEP production for the energy-hungry transition economies of southeast Asia. The environment, both human and physical, is irrevocably changed by dams such as India and Bhutan's Tala dam project, and this causes conflicts, particularly with conservationists. More indirectly, the underlying anthropogenic causes of current climate change are raising temperatures and trigger more extreme events like glacial lake outburst floods, which in turn threaten glacial systems from Antarctica to the Arctic, and from the Himalayas to the Rwenzori Mountains of East Africa.

The Rwenzoris demonstrate the conflicting demands well since, despite being protected by UNESCO's World Heritage and local National Park status, the pressures upon them are increasing, ironically more from outside the region than within it.

This is obviously similar to Antarctica, although here, with no indigenous population, conflicting demands stem from its resource value to outside players. These include conservationists like Greenpeace and research organisations like the BAS, wealthy European and American tourists and Japanese factory fish fleets.

So far the Antarctic Treaty and its various annexes have held at bay any fossil fuel resource extraction, whereas in the Arctic, BP and Statoil are examples of TNCs currently prospecting for oil and gas resources as the global energy crisis and 'peak oil' fears grow. The 30-year-old debate about drilling for oil in the Arctic National Wildlife Refuge is another example of conservation meeting exploitation head on in a periglacial environment famous for its tundra and boreal forest systems. However, the change in the main gatekeeper player here — the US government moving from the Bush to the Obama administration — means this cold environment may have less direct pressure.

It is obvious from this discussion that there is great complexity in the places, the people involved, and the power wielded by gatekeeper organisations. The huge variety of cold environments across the world have a common denominator: they have fragile physical systems easily disrupted by demands of whatever nature. Both active and relict glacial and periglacial environments have long been important as a water, power, mineral and tourist resource; but the added demands of the twenty-first century for conservation of physical, human and aesthetic value means increasing conflicts. The latest threat of climate change will put extra pressure and may even increase demands for water and low carbon energy production, especially on actively glaciated areas, which effectively act as a global regulator of Earth systems. There is no doubt about the increasing nature of pressures despite any mitigating management.

ⓔ This conclusion would be awarded **15/15 marks** for the following reasons:
• It is clearly stated.
• It has thorough recall of content/case studies used in the report.
• It shows understanding of the complexity of the question, and returns to the underlying synoptic context of place, people, power. Having direct and indirect demands is a rewardable, higher-level concept.

Note that the degree of ongoing evaluation within the report, which is also needed for full marks, is not obvious from this extract. You will also see that it has excellent vocabulary and is written in the third not the first person, a better style generally that will help its quality of written communication mark.

# Option 3 **Life on the margins: the food supply problem**

## Question

**To what extent do food security issues vary across the world?**

## Pre-release research focus

- **Explore why food security varies spatially.**
- **Research a range of rural and urban locations experiencing food security issues.**

## Plan

ⓔ Use BUG (see page 90) for a quick way of unpicking the question and setting out your main approach. For a detailed plan use the DRACQ framework (Table 4) — these letters stand for the sections in the mark scheme that examiners use to award marks:

    **D** = discuss/define
    **R** = research
    **A** = application
    **C** = conclusions
    **Q** = quality of written communication

**Table 4** Food security plan

| | |
|---|---|
| D = 10 | Definition — food security includes food supply, use FAO |
| | Large topic — say going to focus on certain aspects: famine and drought/desertification |
| | **Yes** variations — desertification. Overnutrition New York and undernutrition Mumbai |
| | **No** similarities — famine human factors politics Zimbabwe and Korea, drought Australia and Ethiopia |
| R = 15 | Food security issues |
| A = 20 | Basic issues = feast–famine i.e. overnutrition and undernutrition |
| | Sub issues = increased globalisation of production and consumption |
| | Spatial: differences between urban/rural and different economic groupings: MEDC–NIC–LEDC–LDC spectrum |
| | Management — mitigation reducing issues 2009 Task Force UN |
| C = 15 | Evaluate after each case study |
| | Return to all at end. Increasing issues globally? |
| Q =10 | Vocabulary sections |

For this option we will look at a sample introduction and a sample case study.

# Student answers

## Introduction

Food security is defined as the 'reliability of food supply'. Globally an estimated 850 million people are chronically hungry and up to 2 billion people lack food security intermittently (*Edexcel* A2 *Geography*, Dunn et al.). According to the FAO there is enough food per person (using an average of about 2,700 calories of food per day). But food is not produced or distributed equally. Some areas produce much more food than others, with a range of factors involved — see the diagram below.

Factors affecting food security

Access to food is still seen by many as a privilege rather than a basic human right — hunger causes the deaths of about 35,000 people daily. Malnutrition is even more widespread as an issue. Surprisingly this affects richer economies too, such as Canada where 2.5 million people depend on food banks (International Development Research Centre Website), and an estimated 30 million people in the USA may be unable to buy enough food to maintain good health. So although it might be thought that only rural areas suffering desertification, like Ethiopia, have poor food security, in fact a growing issue is that of food security by the urban poor in expanding megacities such as Mumbai and even in cities in more developed economies like New York.

In recent decades, health and nutritional status have improved considerably, but are still far from satisfactory in many countries. Ironically in many richer economies the opposite issue is emerging: that of obesity — malnutrition of a different kind.

The rising middle classes of transition economies like China are generating another issue: that of changing food habits with increasingly Western preferences for meat rather than traditional types of food. This is putting pressure on rural areas hundreds of miles away in our globalised world, as soya is grown for external markets for cattle fodder rather than crops being grown for local consumption. Meanwhile, food issues in countries like the UK centre around food miles, ecological footprints, cheapness of foods, production methods like animal welfare and organic products as well as obesity.

The earliest known famine was in 3500 BC in Egypt, and famine has affected places as far apart as China, India and Ireland. The worst famine to date was probably in nineteenth century China, killing up to 13 million. The FAO states that such major disasters increased fourfold between the 1960s and 1980s, but the international donor community has become better at preventing them turning into catastrophes. Recently there have been extra efforts to address famine in places like Ethiopia by NGOs such as Comic Relief. Red Nose day was started in response to the 1985 famine there. In the LDC of Kenya in 2009 the president declared a national food shortage, showing that years of effort trying to improve food security have not been enough.

The fact that the UN set up a Task Force on Global Food Security in 2008 indicates that food security issues are far from sorted. The Task Force includes many UN agencies like WFP, WHO, UNICEF and the World Bank.

**e** This is a good discussion on a general range of food security issues, with relevant examples. However, it does not signpost clearly to the examiner the framework to be taken, either in terms of concepts, e.g. which food issues, or specific case studies, e.g. China, Mumbai.

In the plan there was an indication that food security issues common to different areas and different in locations would be assessed, but this simple and effective two-fold division to help argue points does not appear clearly in the introduction.

It would be awarded **8/10 marks** but could have achieved top marks with the addition of an extra paragraph to justify and illustrate the chosen issues, as follows:

This report will focus on the following food security issues…and case studies to assess how the issues of food security vary across the world and also over time:
- the NIC megacity Mumbai because…
- national scale: RIC of China, which shows…
- current insecurity issues in Zimbabwe, especially in the capital Harare, because…

| Student B |
| --- |

## *Factfile on China (sources: FAO, CIA, BBC)*

China is a good example to show how food security varies spatially within a country, and also over time. China feeds over one-fifth of the world's population with only one-fifteenth of the world's arable land. Access to food is governed by poverty and although national economic growth is strong, it is uneven across regions. Farmers' incomes in the central and eastern regions of China continue to grow more rapidly than those in the west and southwest. Income inequality among regions, between rural and urban areas, and within regions continues to grow. More challenges of food supply are created by degraded land, water and resources, and increasing cost of labour and domestic food production. In 1958 and 1961 China had widespread famines with an estimated 15 million associated deaths. The Nobel prize winning economist and expert on famines Amartya Sen says that most famines do not result just from lower food production, but also from an inappropriate or inefficient distribution of the food, often compounded by lack of information and indeed misinformation as to the extent of the problem. In China, the reasons have been:

- largely from a series of natural disasters, floods and droughts especially, compounded by poor management
- the huge institutional and policy changes from the 'Great Leap forward ' policy involving forced collectivisation, which reduced the incentives for peasant farmers to work effectively; labourers were often forced into working in manufacturing instead of food supply

The famines were associated with land degradation, which has had more effective management to resolve it as shown in the diagram below.

| Increased use of more **marginal land** for arable production: reclamation of wetlands, forest, semi arid land. | **Land degradation:** Deforestation, desertification, soil erosion, river silting, dam life shortened etc & reduced biodiversity. **All these = reduced food security** Lack of sustainable use, reduced production | **Grain for Green Policy** 1999+ started in Sichuan, Shanxi & Gansu, since spread. 7m ha low yield land returned to forest & pasture 1999–2003. Less marginal land used more effectively & more imports |

Data on China from the NGO 'Leadership for Environment and Development' established by The Rockefeller Foundation

However, despite the pessimistic forecasts made in the 1960s, famine has not been a serious issue in China in the past 50 years because of land reforms and the development of a market economy allowing wealth to spread. In fact about 20% of the 1 billion overweight or obese people in the world are Chinese and China is fast catching up with Western levels of obesity; 12% of adults in cities are now classed as obese. The speed of the nutrition transition from a low-fat to a high-fat diet is one of the highest seen globally.

Sub-conclusion: this case study shows both spatial and temporal differences in food security, and a shift from famine and land degradation as the main food issues to changing food tastes and health issues linked to obesity.

e The main report section shows one well-selected case study that addresses many of the complex points involved in the question. The use of statistics and referenced data combined with an evaluative sub-conclusion ensures top marks will be awarded for selection, application and conclusions. There is also excellent geographical terminology, for example the words degradation, nutrition and transition.

# Option 4 **The world of cultural diversity**

## Question
'Cultural attitudes determine how landscapes and environments are valued.' Discuss.

# Pre-release research focus
- **Explore the attitudes of different cultures to the environment and landscape.**
- **Research a range of landscapes and environments to see how these attitudes may affect their value as shown by management and use.**

## Unpicking the research focus
**e** Identify a range of different cultures and attitudes and uses to/of environment/landscape. Use the examples/case studies from Tables 5–7, which can be used or rejected in the final exam.

**Table 5** Environment: different attitudes and cultures

| Cultures with differing attitudes to the environment and landscape | | Cultures with similar attitudes | Cultures with a change in attitudes over time |
|---|---|---|---|
| **Exploitation focus**<br>USSR: Aral Sea, Chernobyl<br>China (see changes) | **Protection focus**<br>UK<br>Hindu and Islamic religions<br>Ethnoscapes: Aborigines Uluru sacred site<br>Global: UNESCO and World Heritage Sites, e.g. Marrakech traditional market (souk) and Djemaa el Fna, Peru Macchu Picchu, Jurassic coast UK | The global green and conservation movement: a culture not bound by national frontiers<br>e.g. Greenpeace, WWF, FoE and green political parties, e.g. German or Kenyan Green Party<br>Global corporations/TNCs and architects and planners: cloning of urban financescapes<br>National Trust NGO and 2009 strategy on sustainability and cultural preservation | UK — past Industrial Revolution and exploitation turned to pro environment after 1960s especially. New National Parks and Conservation Areas in cities.<br>USA Wildscapes and ecoscapes and Obama on global warming Public attitudes to climate change<br>Changing attitudes to wolves<br>China possibly recently: Beijing Clearskies policy, e.g. new National Parks<br>Rebranded landscapes e.g. coastal resort of Boscombe, Olympic site in London, cities with 24-hour culture, e.g. Birmingham |

**Table 6** Some types of use and management

| Types of use | Types of management |
|---|---|
| No use, preservation<br>Use for resources under surface e.g. oil, coal<br>Use of ecological resources, e.g. forest products, water abstraction<br>Use for urban demands: ecological footprints<br>Use for tourism: mass, eco, elite | Top down: often government-led large schemes<br>Bottom up: grassroots movement to preserve/protect/conserve<br>Management from pressure groups and NGOs<br>Sustainable management: long-term planning for future generations |

**Table 7** Factors affecting how landscapes and environments are managed

| Social/cultural | Economic | Physical |
|---|---|---|
| Strength and commitment of governance<br>Role of IGOs like UNESCO or UNEP<br>Role of individuals, freedom and apathy<br>Religious/ethical attitudes<br>Consumerism and McDonaldisation | Finances available to protect/preserve/find alternatives<br>Role of TNCs, e.g. in polluter pays<br>NGO and Foundation funding<br>Poverty and wealth | Scale of area or environmental problem<br>Technology available to use<br>Surveillance to check management working |

## Plan

**BUG** the question (see page 90 for more on BUG).

'**Cultural attitudes** determine how **landscapes** and **environments** are valued.' Discuss.

**Student answer**

## Introduction

There is an infinite mosaic of natural, semi-natural and human-dominated landscapes across the world, from the wildest landscape of Antarctica to the middle of downtown New York. As for the environment, this is made up of the atmosphere, hydrosphere and ecosphere as well as the land. Humans are part of the natural Earth systems, or so some viewpoints would say, since other viewpoints see humans as a separate entity with the Earth as its resource. This is the fundamental reason why there are differing attitudes, plus the fact there are so many players involved: from international ones like UNESCO and UNEP to individual governments like the UK and India, trade blocs like the EU, businesses and TNCs like Shell, green pressure groups like Greenpeace and WWF and last but not least, individual attitudes. An urbanised viewpoint is increasingly dominating attitudes, since over 50% of the population live in the megacities and world cities of the world from Beijing to London. The diagram below summarises these attitudes and the key case studies that will be used to discuss them.

| Mainly exploitative attitude today | Used to be exploitative but more protective recently | Used to be protective but now more exploitative | Mainly protective attitude today |
|---|---|---|---|
| Westernised, humanist society<br>Russia | UK National Parks and sustainability<br>Rebranded historic/cultural landscapes, e.g. Boscombe, Bradford, Stratford, Olympics | Hindu faith, India | Romantic and ecological viewpoints<br>GAIA concept<br>WWF<br>Greenpeace<br>UK USA Environmental Agencies |

Attitudes and key case studies

## Section 1 An overview of attitudes with their historical and religious links

### 1.1 Ecological and exploitative views summary

Different people and groups attach different values to the same landscape or part of the environment. For example, trees may be valued for aesthetic, ecological, food and profit reasons depending on people's current wealth and general culture. Similarly, wolves and tigers, once seen as a danger, are now seen as key species by ecologists and an embodiment of 'wildness'. This introduces the concept of changes over time in our attitudes.

### 1.2 Roles of religion and technology

Long-established religions have usually viewed the environment as sacred, with every tree, river, animal and bird containing a spirit and needing reverence. Islam, Buddhism and Hindu faiths are good examples of this. The more recent, allied viewpoint is the Gaian attitude to Earth as one system. Professor Lovelock stated in 2004 that global warming is 'the response of our outraged planet'. Advocates of this view on the environment want to use technology like renewable energy and even nuclear power to manage the environmental repercussions of human-led change.

However, the attitude fostered by Christian faiths over centuries is an anthropogenic standpoint where nature is a type of utility to humans and traditionally exploited on an ever-increasing scale as technology, population pressure and consumerism have developed. Environmental degradation was so obvious by the 1960s that the Green movement was created, originally in the USA. Here, the backlash from the extreme use of the landscape, where humans were technologically dominant, combined with freedom of speech and the new perception of Earth resulting from satellite imagery and space travel, helped form some new attitudes:

- The Romantic–Preservation ethic: where nature is a type of god, vulnerable to human activity, and should be preserved or protected as much as possible. The priority use is therefore nature reserves, parks and wilderness areas — policies used by many NGOs and governments today. Unfortunately many of these can be visited only by wealthy elite tourists, usually from the capitalist culture, and those attracting most visitors may even be 'loved to death', as seen in the Lake District and Yosemite National Parks when honeypot areas become super saturated with visitors. It is an attitude that is almost voyeuristic, and any indigenous people may suffer in core areas where pristine 'nature' is designated. This has led to well-known culture conflicts, for example at the sacred site of Uluru in Australia between the native Aborigines and the government. The WWF has even expelled people from some reserves, as in Korup.
- The Resource Conservation ethic focuses on the greatest use of the resources of an area for the greatest number of people for the longest time. Exploitation of resources is reduced or not allowed. This has sustainability at its core, and is followed by most Westernised governments since the 1992 Rio Summit. This includes the concepts of Polluter Pays, Precautionary Principle, Local Agenda 21, and most recently contraction and convergence as part of tackling the pressing issues associated with climate change management. It also tries to involve all the stakeholders in an area and create a working relationship between nature and people.

## 1.3 Russia

However, not all cultures considered as 'Westernised' have the same change in approach to the environment as the UK, Europe and the USA. The table below is a timeline of Russian attitudes to the environment, taken from the Russian branch of Greenpeace and compiled by Tsyplyonkov: Russia has transformed itself from a capitalist to Marxist to hybrid social type culture in the past century. Today it is infamous for continuing its exploitative attitude to the environment — both at home and in its ecological footprint: 'the Russian government is prepared to pay the ecological price for economic development'. This change in attitude is also being seen in China and India, as the following case studies of countries in economic transition show.

| | Twentieth century up to 1990 | Generations of Russians have grown up with the attitude reinforced by their government, often forcibly, that the state is dominant and individuals are not responsible for environmental care. This shows in low rates of responsible waste disposal. The explosion at the Chernobyl reactor in Soviet Ukraine in 1986 was the world's worst nuclear disaster. The late twentieth century saw grandiose schemes such as diverting rivers, as in the Aral Sea tragedy, and building of many nuclear power plants including Chernobyl and polluting industries like around Lake Baikal. |
|---|---|---|
| | 1990s–2000 | The huge change in political culture in the 1990s with the collapse of the Soviet Union created a short-lived change in attitude with the formation of an Environment Ministry, but by 2000 this was abolished. Separate departments mean no integrated overview of the environment exists, nor any one agency has ultimate responsibility. This contrasts with the situation in most European countries and the USA, which have overseeing Environment Agencies. |
| | 2004 | Russia signed up to the Kyoto agreement in 2004 only after pressure from the EU linked with trading relationships. |
| | 2009 | The basic attitude in Russia seems to be that environmental restrictions pose a threat to economic development, seen most recently in the huge environmental impacts of transforming the Black Sea resort of Sochi for the 2014 Winter Olympic Games. |
| | 2007 | The task of changing individuals' perception of the environment is being led by international green groups like Greenpeace and WWF. Public outcry against a proposal to build an oil pipeline close to Lake Baikal helped persuade the government to stop its development. |

Timeline of Russian attitudes to the environment

## 1.4 China and India

China has immense environmental issues, as highlighted by the name and shame website the Blacksmith Institute. However, there is the start of a shift here as the government wants to display a more developed world attitude to environmental issues, demonstrated in the 2008 Beijing Olympic clean-up and its Blue Skies policy.

In India, the original Hindu attitude to environmental protection has been lost in the drive towards economic progress, with deforestation in the Himalayas, desertification in Rajasthan, and pollution in cities, for example, pollution of the River Yamuna in Delhi and air-pollution in Chennai. The Indian environment-protection movement that opposes this degradation of natural, farmed and urban landscapes is organised largely by Westernised elites and based on Western models. It has failed to become a mass movement because it lacks spiritual foundation and content necessary to inspire Hindus (quote from the philosopher Professor David Frawley).

## 1.5 Zoning and mixing

The concept of *himil*, meaning protection of certain zones, has long been used by the Bedouin tribes as a custom/tradition inherited from their ancestors, and can still be seen in many Muslim countries, such as Saudi Arabia, where it is practised by the government to protect wildlife, e.g. efforts to save the Arabian oryx from extinction. Islam, as a way of life, expects humans to conserve the environment. It is not a culture but a body of principles and values, which may be assimilated into whatever the local culture is. Cultural landscapes in many larger cities across the Westernised world now feature components of such mixing: from mosques to supermarkets, and even schools such as Islamia in London, which was originally started by former pop star Cat Stevens, now known as Yusuf Islam, and is now funded by the British government.

## 1.6 Cityscapes

Heritage landscapes in cities are increasingly valued as a commodity, as cultural tourism helps rebrand places as different as Boscombe and Bradford. Efforts to keep individuality of places may be seen in the 'home town–clone town' debate. Globalised forms of consumerism in retail high streets and especially financescapes are often fought against to give places a special place value and resist cultural imperialism. That is why places like historic Winchester may be more valued for some elite footloose businesses than say the more utilitarian cityscape of Basingstoke, although even here the cityscape is dominated by TNC land uses like Starbucks and McDonald's. The ultimate globalised cityscape is probably the large-scale shopping mall, where attitudes to consumerist culture and domination by some large TNCs mean a very standard layout and look from Putrajaya in Malaysia to Shanghai in China, Westfield in London and of course where it all started — downtown malls in any US city. In London the new Olympic landscape has strong overtones of environmental protection and eco-friendly/low-carbon buildings and landscapes, although the original urban landscape was partly destroyed despite local opposition in Stratford.

## 1.7 Sub-cultures

Sub-cultures may have an effect on landscapes too: whether rural such as the agriculturally based Mennonites in Belize or urban such as the hip-hop culture of graffiti and street culture.

Lastly, this report will discuss some of the key players other than pressure groups and governments.

## 1.8 Players

At the lowest scale, but by far the most numerous at over 6 billion, is the individual. Attitudes to landscape management and environmental issues vary, as shown by the 2007 HSBC survey of public attitudes towards climate change in nine countries across the world, involving 9,000 people. The HSBC Climate Confidence Index shows that people in developing economies such as China and India exhibit the greatest concern, commitment and optimism towards the problem of climate change. The greatest indifference, reluctance and fatalism is evident in developed economies, especially France, the UK and Germany. Across all countries, 68% thought governments should be playing the leading role, rather than NGOs, companies and individuals.

UNESCO is a key player in fostering cultural diversity and preserving diverse cultural landscapes globally, being the largest international organisation to date. Countries sign up to its ideals, including the UK and the EU bloc. It creates protected areas and protects buildings of global significance.

The EU has a range of policies designed to preserve and enhance the landscape and environment, from the Regional Development Fund for heritage restoration in Ebbw Vale (an ex mining and steel town in south Wales) to the LEADER and LIFE III projects (source: Europa.eu). Many of these restore landscapes, if not to an original form at least one that is less polluted and more visually pleasing, and can even become a resource: as in Swansea's 5 Parks scheme in an area once the worst degraded in Europe. This shows the ideas of Carl Sauer back in the early twentieth century, who thought that cultures develop because of the landscape but also mould the landscape as well.

At a country scale, the UK has a sustainable communities policy and many others designed to respect and improve the environment, ranging from National Parks to recycling targets, renewable energy plants to conservation areas in cities.

## Conclusion

There are many complex, interrelated factors influencing attitudes to landscape and the general environment, many based on culture in its widest sense. It has been shown how religion is a major cultural influence in attitudes, but above all the role of economic development outweighs environmental costs, especially in Russia. The present backlash focusing on environmental climate change may be too late with the tipping point described by Stern and Cox as imminent or even passed. It has been shown how international efforts are involved to both create cultural diversity and environmental protection, with NGOs like Greenpeace and IGOs including UNESCO as major players, working in conjunction with organisations like the EU and nation states. Landscapes include both wildscapes and cityscapes, and there are efforts to preserve diversity in landscapes both for philanthropic reasons and to increase their commodity value as tourism increases.

ⓔ This is a very good report, which would score around 60 marks (an A grade). What makes this an excellent report?

- It is well referenced, from a range of reputable sources.
- It has a range of case studies, some in considerable detail, which enable an in-depth discussion.
- It covers differing scales: global, national and local.
- It uses vocabulary in a sensible, knowledgeable way.
- It relates each example/case study back to the question, with sub-conclusions at the end of each section.
- It mentions some theories effectively without getting carried away on a critique of them.
- It has a distinct report style but a flowing prose discussion.

How could it have been improved? A slightly longer conclusion, returning to most of the main points/examples would have helped. The conclusion also introduces a few new points not covered in the report. The student could have included a clearer methodology.

# Option 5 **Pollution and human health at risk**

## Question
**To what extent do health risk causes and patterns vary?**

## Pre-release research focus

- **Explore the complex causes of health risk, and any links to socioeconomic status and geographical features.**
- **Research a range of health risks and how models may help the understanding of their patterns and trends.**

### Unpicking the research focus

**ⓔ What is a health risk?** Use classification into infectious, chronic, pollution/toxic and traumas.

**Causes and patterns** suggests a focus on enquiry question 2: what are the causes of health risk? Other enquiry questions may help — for example, enquiry question 4 on management to mitigate causes and impacts. Start with a simple summary of a range of causes (Table 8).

**Health risks** include infectious, chronic and trauma risks, so you need examples of each, linked to epidemiology and health transition models, for example China, the UK.

**Table 8** Range of causes

| Human causes | Physical causes |
|---|---|
| Lifestyle choices: exercise and diet <br> Vulnerability, depending on: <br> • Wealth — employment and income: socioeconomic status <br> • External factors of health service: state/private/ insurance <br> • Role of management and effectiveness by international/ national/local authorities and NGOs and businesses and charities | Environment: geographical features — disease reservoirs, vectors <br> Changes in environment: climate change, e.g. shift of malaria <br> Housing: displaced people, refugee camps, e.g. Darfur <br> Pollution: Kuznet curve suggests differences between LEDCs/NICs and MEDCs |

## Plan

**BUG** the question (see page 90 for more on BUG).

**To what extent do health risk <u>causes</u> and <u>patterns vary?</u>**

You can do some quick planning around the question as shown in Figure 4:

Have to weigh up the variations, see if any are similar, e.g. infectious disease of HIV and measles — Switzerland, USA, UK
Extent of differences — scale different in…
By end have to decide

Human causes, i.e. lifestyle choices by individuals, wealth, technology, management by WHO, Govt, NGOs, overuse antibiotics in MEDCs…

Physical causes, e.g. — local environmental conditions — river pollution, global warming

**To what extent** do health risk causes and patterns vary?

Show range and complexity of health risks: infectious/communicable diseases, e.g. measles, malaria
Chronic/degenerative diseases, e.g. obesity related diabetes & depression
Toxic risk from pollution — cancer villages China
Trauma — industrial accidents sweatshops

Spatial pattern:
• global, i.e. MEDC–NIC–LEDC–LDC spectrum
• local, e.g. within a city — e.g. Manchester
Temporal pattern–epidemiological model
Models may help even though not specifically asked for: contagious diffusion, measles and hierarchical diffusion — HIV

**Figure 4** Plan

The following answer is the outline for the main analysis section, given marks in the Research and Application part of the mark scheme.

**Student answer**

## Section 2.1: Infectious diseases: the case of measles, a long-standing preventable and curable infection

### Overview
According to the UN office for the coordination of Humanitarian Affairs, by 2008 the Global Measles Initiative had helped the vaccination of over 400 million children in 50 countries. The players involved are the American Red Cross, the US Centres for Disease Control and Prevention, the UN Foundation, UNICEF and WHO. They aim to reduce measles deaths by more than 68% globally and 91% in Africa compared with 2000 rates.

### Case study 1: Pakistan
Pakistan is currently attempting to vaccinate 64 million children in just 1 year in the effort to eradicate this highly infectious disease, which can cause complications of pneumonia and encephalitis and a high mortality rate. Being only at stage 2 of the epidemiological model, it is still dominated by acute infectious disease, largely because it is a low-income economy (GNI of US$870). However, according to the Population Reference Bureau website, Pakistan is entering a new phase of increased chronic health risks because of its rapidly ageing population, so it will have a double burden to cope with.

## Case study 2: UK

In comparison, the health risk from measles in the UK (GNI US$42,740) is minimal because of mass vaccinations in the late twentieth century. However, since 2004 there have been outbreaks, mainly because some families have chosen not to immunise their children owing to an unfounded scare over the MMR injection.

## Case study 3: San Diego

The pattern of disease follows the contagious diffusion model of Cliff, where direct contact is needed between hosts of the disease. This was shown by the 2008 outbreak in San Diego USA caused by a group returning from holiday that had had contact with an infected person in Switzerland. Despite being a highly developed economy, which is really at stage 3 of the epidemiological model, Switzerland has no centrally structured vaccination plan. This particular outbreak travelled the world in less than a month because of the air travel factor. The San Diego family had refused the vaccine, and this was the first time since 1991 the city had seen measles.

Sub evaluation: The causes are basically the same, i.e. the presence of the pathogen, but it is the management that makes the health risk so different in these countries — poverty/wealth and the efficiency of the health system influencing the vulnerability of the population. The scale of the problem in Pakistan obviously needs outside assistance. Yet it comes down to individual lifestyle choices as well.

## Section 2.2: The case of an emergent, preventable but incurable infective disease: HIV

Unlike measles, the newer emergent infectious disease of HIV has no cure, but with proper treatment and nutrition, mortality may be delayed. HIV is a problem for differing groups of people, i.e. its pattern differs markedly both globally and more locally.

**e** Here the student wrote a discussion of the different strains and patterns of HIV, groups of people most affected, role of antiretroviral drug campaigns, RED, Melinda and Bill Gates Foundation aid and the WHO HIV/AIDs management strategies…

## Section 2.3

This was followed by an evaluation of the variation in chronic diseases, which are now creating a double burden especially in transition economies like China…

## Section 2.4

And still on China, a section follows on pollution as a major cause of toxic health risk, using the Blacksmith Institute and Reuters news agency as sources. The focus is on cancer clusters like the village of Huangmenying in central Henan province, poisoned by unregulated effluent from tanneries, paper mills and an MSG factory flowing into the main water supply of the Huai River. The causes are linked with ineffective environmental quality standards and health cover in the country. The pattern is that all age groups are affected but in a relatively small hot spot of pollution — although as a nation China is now infamous for its poor environmental quality and under-resourced healthcare system…

### Section 2.5: Environmental change at a global scale: climate change and the potential rise of malaria

Discussion of past, present and future scenarios for malaria, classed as an endemic re-emergent disease in especially Africa (e.g. Nigeria), owing to a complex mix of causes: chloroquine resistance, civil unrest and migrations, increased rainfall, dam construction, reduced health budgets and drug availability, high birth rates producing vulnerable under 5 year olds… [sources: Wellcome Trust, WHO, UK government DFID.]

**ℯ** This section would score a maximum 20 marks for the analysis and application section. Why?
- It is well referenced, from a range of reputable sources.
- It has a range of case studies, some in considerable depth, which enabled a decision on the command word in the final conclusion.
- It covers differing scales: global, national and local, although it does not manage to cover the city scale, e.g. research on Manchester's differing morbidity and mortality patterns.
- It uses vocabulary in a sensible, knowledgeable way.
- It relates each example/case study back to the question, with sub-conclusions at the end of each section, so will gain marks not just in research and application but also in the conclusions section.
- It mentions models effectively without getting carried away on a critique of them.
- It has a distinct report style but a flowing prose discussion.

How could it have been improved? A quick sketch of the diffusion model, annotated to show the examples chosen, or the Kuznet curve, which appeared in the student's research focus but was omitted in the exam, would have helped.

# Option 6 **Consuming the rural landscape: leisure and tourism**

## Question

**'The use of rural landscapes for leisure and tourism is more controversial in some areas than others.'**
**Discuss.**

## Pre-release research focus

- **Explore the impacts of leisure and tourism as a consumer of rural landscapes, and the role of management.**
- **Research a range of locations showing differing impacts from recreational and tourism users.**

## Unpicking the research focus

ⓔ Identify the key areas of the research focus — sometimes a simple bullet list is all you need:

• How leisure and tourism = a consumer.
• The users. Who is involved in the use and management of rural areas — people/players: stakeholders, gatekeepers within area and outside — increasing role of international organisations from pressure groups, conservation to TNCs exploiting rural areas and UN, plus individuals.
• The uses. What are the uses: types of leisure and tourism, including comparison of overlaps between leisure and tourism activities.
• Elite, eco, mass.
• Active, passive tourism etc.
• Uses and impacts: positive, negative, direct, indirect, short and long term, economic/social/environmental etc.
• Role of management in increasing/decreasing impacts.

    Create a series of factfiles on places showing these different impacts. You could categorise these into a spectrum diagram (Figure 5).

| Low impact at present | Was high now lower because of effective management | High impact at present |
|---|---|---|
| Antarctica | Macchu Picchu | 2nd homes Snowdonia, Caribbean e.g. Barbados |
| Wolong | New Forest | |
| S Downs | Mallorca | |

**Figure 5** Spectrum diagram

## Plan

**BUG** the question (see page 90 for more on BUG).

    **'The use of rural landscapes for leisure and tourism is more <u>controversial</u> in some <u>areas</u> than others.' Discuss.**

Sort out your arguments — a simple table (Table 9) can help.

**Table 9** Controversial or not?

| Yes — very controversial because | No — less controversial because |
|---|---|
| High negative impact — Snowdonia second homes, loss of farmland and tropical rainforest St Lucia<br><br>Changing leisure use e.g. mountain bikes Glen Shee Cairngorms<br><br>Increasing impact — Antarctica, Kilimanjaro<br><br>Less effective management — golf courses Vietnam | Less impact — remote<br><br>Better management — core zone Yosemite, Annapurna — shows TEMPORAL CHANGE<br><br>Partial management — fox/drag hunting Hampshire |

The following answer shows how one student approached the introduction of this report. Read it carefully, consider the examiner's comments at the end and think about how the answer could be further improved.

**Student answer**

## Introduction

Rural landscapes are complex to define but they are typically classified as less populated parts of the planet with low population densities, and range from wilderness to urban fringe areas.

Leisure and tourism are features of economies globally nowadays and both have been increasing, especially in transition economies such as China.

The question requires an understanding of what is meant by controversy and all that entails, especially conflicts between different users or players. The essay will also discuss how the use of the rural landscape is on the rise and how that too can cause challenges, especially where there is intensive use and the carrying capacity is exceeded. However, management by local, national and international players may reduce the negative impacts.

I will use a variety of case studies to illustrate my answer, including the debate surrounding increased use of Antarctica for ecotourism and the globalisation of golf in, for example, Vietnam. I will also cover smaller-scale examples of conflicts that I have researched, including the use of motocross in Devon (post-productive landscape) and overcrowding of urban country parks. Other controversial issues such as fox-hunting bans in the UK, together with the 2009 Comic Relief Kilimanjaro climb, will be evaluated.

ⓔ This is a good introduction, scoring **8/8 marks**.
- There is clear reference to the question with the development of a focus.
- There is a section on the framework to be used according to both context (small, large, past and current) and places (e.g. Antarctica, Kilimanjaro).
- There is less accurate definition of key terms: leisure, tourism, but a good comment on the rise of these activities.

It could have been improved by a link to negative impacts, which create controversies. It should have been written in the third person, not the first person.

## Knowledge check answers

**1** Usually vulnerability decreases with increasing development. Magnitude increases vulnerability, but has no bearing on development.

**2** Quick and easy to construct; give a good visual representation; useful for making comparisons between different types of event or events at different times.

**3** Japanese harbour waves — frequent events in that country. Seismic sea waves is a more technical term.

**4** Different viscosities will have big impacts on the nature of the eruption, i.e. Hawaiian (fountains of lava) vs Strombolian vs Vulcanian vs Plinian (high viscosity dacitic or rhyolitic).

**5** Data from World Disaster Report (2010) — data 2008–2009: Earthquake deaths (including tsunami) = 453,000, volcanoes 230, droughts 230,000, floods 53,000.

**6** A trend is a change in a set of data that has a direction, i.e. developing or changing: up or down etc. Trends might be described as rapid or slow, steep or shallow.

**7** Players can be called interested parties or stakeholders. Remember they can be groups, organisations or individuals.

**8** It helps you to understand how a theoretical chain of events should unfold following a disaster; it allows you to link this to an actual (researched) sequence of events in terms of recovery and evaluate them against the model. All models are generally flawed since they simplify real-world systems and processes.

**9** Risk is the potential that a chosen action or activity (e.g. tectonic event) will lead to a loss. Vulnerability is how predisposed (or disadvantaged) an individual or population is to a particular event — see Table 6, page 17.

**10** Inputs are parts going into a system, e.g. snow, weathered rock. Processes are events happening to transform the systems, e.g. erosion and deposition. Outputs are parts coming out the other end, e.g. meltwater, deposited till.

**11** The highest level to which the fresh snow on a glacier's surface retreats during the melting season. It is a line separating the accumulation area from the ablation area.

**12** Paleoclimatology is the study of past climates — often climate proxies are used: microbial life (diatoms, forams), also tree rings and sediment cores, which include pollen.

**13** The glacial maximum refers to a period in the Earth's climate history when ice sheets were at their maximum extension, estimated to be between 26,500 and 19,000–20,000 years ago, marking the peak of the last glacial period (Pleistocene).

**14** All descriptions of ways in which the Earth's atmosphere prevents all of the insolation from the Sun reaching the ground surface. All three are important for controlling the Earth's heat energy budget.

**15** This is commonly called the North Atlantic Drift or Conveyor. The oceans are very important for redistributing heat energy from within the sea. In the case of the North Atlantic, warm water is moved northeast from the equator, while cooler water is returned southwest.

**16** Periglacial refers to places on the edges of glacial areas ('peripheral'). These areas were not buried by glacial ice, but were subject to intense freezing cycles. Weathering is the mechanical and chemical breakdown of rock and mineral matter *in situ*.

**17** Malnutrition results from a poor diet, when some nutrients are either lacking, taken in excess or in the wrong proportions.

**18** The FAO works to defeat hunger, by developing better farming, forestry and fishing practices worldwide.

**19** Examples could include supermarkets such as Tesco and Asda (Wal-Mart), which sell food, major food suppliers such as Del Monte and Cargill or fast-food TNCs such as McDonald's.

**20** People are considered food secure when food is available, affordable and utilised so people can be healthy.

**21** Food miles refer to the distance food travels from producer to consumer. For instance cocoa beans for chocolate consumed in the UK are likely to have come from Ivory Coast or Ghana in West Africa.

**22** The Green Revolution is the development of high-yielding varieties (HYVs) of crops such as wheat and rice, which when combined with fertilisers and modern farming methods, dramatically increased food supply from the 1960s, especially in Asia.

**23** Examples include artificial fertilisers, pesticides and herbicides, farm machinery such as tractors and combine harvesters as well as Green Revolution crop breeding and genetic modification of crops.

**24** Chemicals are used to increase yields, either directly by adding nutrients (fertilisers) or indirectly by reducing loss and wastage (pesticides, herbicides, antibiotics).

**25** Drylands are arid and semi-arid areas, where water supply is limited i.e. precipitation is low, sporadic and often unreliable. Dryland ecosystems are fragile and relatively unproductive.

**26** Desertification has many contributing factors, rather than a single cause. These include physical factors such as natural climatic cycles but also human factors such as overgrazing, poor farming practice, global warming, deforestation and over-population.

**27** Salinisation occurs when irrigation water on farmland evaporates in a hot climate, leaving behind salt deposits, which contaminate soil and reduce crop yields.

**28** Fair Trade guarantees developing-world farmers a fair price for goods, and so helps provide a decent income. Some Fair Trade profits are often invested in community schemes to improve health, education and water supply.

**29** These people include the homeless, and people on very low incomes such as a small number of pensioners and others who spend a high proportion of their income of food.

**30** Shared values include religion, tradition, behaviour, the idea of what makes up a family unit, art, architecture, design and language.

**31** A culture can adopt some ideas from another culture (assimilation), and ideas can gradually spread from one area to another (diffusion). Sub-cultures can also transform cultures from within.

**32** Globalisation is the increased interconnectedness of the world through trade, transport, communication and the movement of people. The idea of a 'shrinking world' is important to globalisation, making the exotic and unfamiliar increasingly rare.

**33** Counter-urbanisation is the movement of urban people into the countryside; these newcomers to rural areas often bring with them modern urban ideas plus a stereotyped view of the countryside as a rural idyll.

**34** World Heritage Sites are places deemed to be of special cultural and/or physical landscape significance. In 2011 there were over 950 worldwide.

**35** A megacity is any city with a population of over 10 million. World cities are cities with a disproportionate global influence, which is often cultural, such as Los Angeles via the film and media industries. Several cities (London, New York) are both megacities and world cities.

**36** Both tourism and immigration have the potential to 'swamp' a culture with new ideas and influences — some positive, some negative — simply through the volume of newcomers. This could rapidly erode the host culture if not carefully managed.

**37** Globalisation allows cultural features and ideas to spread very easily, such as food, art and music. On the other hand there is a danger that one culture becomes globally dominant, reducing diversity.

**38** One of the key criticisms of Western consumer culture is that the driver for wealth, in order to get the next 'must have' thing, actually seems to make people more stressed and less content. The 'tyranny of choice' makes decision about what to buy an unhappy experience.

**39** The social ecology and deep ecology views both require changes in consumer culture, so that cultures would need to consume (and waste) far less in order to protect the environment and conserve resources.

**40** Wealth gives societies the financial means to protect areas, and the time to enjoy them. In addition, post-industrial societies no longer need to directly exploit their own environment for resources (although through trade, they may exploit environments elsewhere).

**41** Sustainable development focuses on improving quality of life equitably and without needless damage to the environment. Environmental sustainability is really about being 'green' and 'ecofriendly' and is less directly about people.

**42** In 2011, well in excess of 2 billion (1.3 billion people in China, 1.2 billion in India).

**43** The WHO (World Health Organization) is part of the UN. Its role is to work to minimise health and pollution risks to increase life expectancy, health and life chances.

**44** See Table 20; examples of infections include TB, HIV and influenza, whereas degenerative health risks include obesity, heart disease and cancer.

**45** In high-income countries, the majority of risk is from chronic, degenerative diseases whereas in low-income countries infectious diseases are much more common.

**46** No, life expectancy varies by 10 years or more due to lifestyle and income factors.

**47** Stage 4. This is because rapidly spreading viruses are re-emerging as global health risks in our globalised, connected era. These new pandemics tend to have high infection rates but lower death rates than in the past.

**48** The drop resulted from declining health, brought about by falling incomes, rising unemployment and also the collapse of the Russian equivalent of the NHS.

**49** A pathogen is something that causes a disease, such as a virus, bacterium or fungus.

**50** Globalisation is most likely to affect transport routes and air travel, as more and more connections are made around the world. This creates new pathways for disease.

**51** NICS, such as China, Thailand and Mexico. This is because they are rapidly industrialising but have yet to develop proper pollution control systems.

**52** The health divide is the gap between health haves and have-nots, i.e. those at high health risk, versus those with low risk and the money to access healthcare when they need it.

**53** The food transition, in industrialising countries like China, is the rapid shift from a diet dominated by vegetables and grains to one high in meat, fat and sugars. Often this change has occurred in less than a generation.

**54** Health is crucial to life chances and the ability to earn an income and live well; people who suffer from chronic illness cannot work, earn little money and may not be able to feed themselves. This is why the MDGs focus on improving health.

**55** Honeypots deliberately concentrate tourism pressure in one area (by offering facilities, car parking etc.) and therefore may actually reduce the pressure on (and indirectly protect) other nearby rural areas.

**56** Fragility refers to how at risk a location or habitat is to change and damage, e.g. by pollution or changing land use within it. In effect it's a measure of robustness. Landscape sensitivity means the same thing, i.e. sensitivity to change. The word fragility is more commonly used.

**57** GIS (geographical information systems) can be used to record land use remotely and to identify sub-habitats and ecological niches. It can be a very powerful tool to assess damage, e.g. deforestation or even footpath erosion at high resolution. It is very useful for looking at change.

**58** The RSPB is a UK-based charity, with a very large membership of over 1 million people. Its focus is on birds and the wider environment; it maintains over 150 reserves. The IUCN (International Union for the Conservation of Nature) is an international organisation dedicated to finding 'solutions to our most pressing environment and development challenges'. It produces the Red List of our most endangered species.

**59** This is adding value through the sustainable exploitation of a previously under-used local resource, generating wealth and employment, e.g. the Prince of Wales and the Duchy brand.

**60** Natural England is the non-departmental public body of the UK government responsible for ensuring that England's natural environment, including its land, flora and fauna, freshwater and marine environments, geology and soils, are protected and improved. The National Park Authority is the legal body in charge of the UK's National Parks. Management of the Parks is on a devolved basis.

**61** The National Parks and Access to the Countryside Act (1949) established the framework for the creation of National Parks and Areas of Outstanding Natural Beauty in England and Wales, and also addressed public rights of way and access to open land.